T0140218

Women in Engineering and Science

Series Editor
Jill S. Tietjen
Greenwood Village, Colorado, USA

More information about this series at http://www.springer.com/series/15424

Florence D. Hudson
Editor

Women Securing the Future with TIPPSS for IoT

Trust, Identity, Privacy, Protection, Safety, Security for the Internet of Things

 Springer

Editor
Florence D. Hudson
Purchase, NY, USA

ISSN 2509-6427   ISSN 2509-6435   (electronic)
Women in Engineering and Science
ISBN 978-3-030-15707-4   ISBN 978-3-030-15705-0   (eBook)
https://doi.org/10.1007/978-3-030-15705-0

This Springer imprint is published by the registered company Springer Nature Switzerland AG
The registered company address is: Gewerbestrasse 11, 6330 Cham, Switzerland

Preface

The world is becoming increasingly interconnected with technology that bridges the physical and digital worlds. This phenomenon, often referred to as "The Internet of Things" (IoT), can bring great value to the world. Healthcare devices such as insulin pumps can be wirelessly connected to glucose meters to automatically inject insulin to a diabetic patient to improve their quality of life, appliances in a home can be activated from a Smartphone remotely to turn the appliance on or off to increase convenience for the homeowner, sensors in buses and trains in a city can inform an integrated transportation system to provide more efficient connections for a traveler to improve the citizen experience, collaborative emergency services leveraging communications from person to person and with transportation, utility and surveillance systems, can improve public safety.

However, the monumental increase in digital connections to physical devices, due to the wide deployment of IoT technologies, also creates great risk. There is physical, regulatory, legal, financial and reputational risk. As evidenced by attacks related to IoT devices, the key attributes of Trust, Identity, Privacy, Protection, Safety and Security (TIPPSS) need to be assessed and addressed for all IoT applications, devices, processes and services. The goal is to keep humans safe, and our infrastructure secure, while leveraging IoT in new ways.

Today, there are many commercial and personal devices being created without the due diligence to ensure trust and security. Engineers need to ensure the "things" that make up the IoT and the systems they connect to are secure, that the devices or services connecting to a device can be trusted, that the identity of the incoming service request or person can be validated by a trusted authority, that the privacy of the data and the individual is maintained, that the humans and the infrastructure using the device are protected, and that we maintain safety and security. This is called TIPPSS for IoT. The TIPPSS elements are as follows:

Trust: allow only designated people or services to have device or data access
Identity: validate the identity of people, services, and "things"
Privacy: ensure device, personal, and sensitive data are kept private
Protection: protect devices and users from physical, financial, and reputational harm

Safety: provide safety for devices, infrastructure, and people
Security: maintain security of data, devices, institutions, systems, and people

Women are working on various aspects of TIPPSS for IoT, securing our future. The women authors for this book are an impressive group leading the way across the world. These wise, wonderful women of IoT include leaders in technology, cybersecurity, pharmaceuticals, identity and access management, policy, constitutional law, government, trust, privacy, venture capital, blockchain, artificial intelligence, big data and analytics. They work for technology industry leaders like IBM and Cisco, research institutions including CERN in Switzerland, universities including Indiana University, University of California—Berkeley, University of California—Santa Cruz, and Virginia Tech, with prior roles in the White House, NASA and the National Science Foundation. Their university degrees cover a wide array including law, astrophysics, engineering, computer science, accounting, international relations, and political science. We all must work together to secure our future.

We thank the Institute of Electrical and Electronic Engineers (IEEE) for supporting the development of the TIPPSS framework beginning with a workshop to discuss the challenges in end-to-end trust and security technology and policy for the Internet of Things in 2016 [1]. IEEE has further developed awareness of the TIPPSS framework through articles in *IEEE IT Professional Magazine* [2] and *Computer Magazine* [3] in 2018, and in healthcare and clinical IoT initiatives.

TIPPSS is a journey. We need it in our smart cities [4], connected healthcare [3], and across all IoT-enabled systems. If it's connected, it needs to be protected. We are honored to have the women securing the future with TIPPSS for IoT sharing their thoughts and insights in this book, and we look forward to many more of you joining us in this important journey.

Florence D. Hudson
Purchase, NY, USA
Florence.distefano.hudson@gmail.com

References

1. IEEE Standards Association (2016) IEEE Trust and Security Workshop for the Internet of Things. © 2016 IEEE, https://internetinitiative.ieee.org/images/files/events/ieee_end_to_end_trust_meeting_recap_feb17.pdf
2. Hudson FD (2018) Enabling Trust and Security: TIPPSS for IoT. IT Professional 20(2):15–18. © 2018 IEEE, https://doi.org/10.1109/MITP.2018.021921646, https://ieeexplore.ieee.org/document/8338006
3. Hudson F, Clark C (2018) Wearables and Medical Interoperability: The Evolving Frontier. Computer 51(9):86–90. © 2018 IEEE, https://doi.org/10.1109/MC.2018.3620987, https://ieeexplore.ieee.org/document/8481273
4. Hudson FD, Cather M (2017) Creating, analysing and sustaining smarter cities: a systems perspective. TIPPSS - trust, identity, privacy, protection, safety and security for smart cities. (ISBN-10: 1848902093, ISBN-13: 978-1848902091)

Contents

1 IoT: Is It a Digital Highway to Security Attacks?   1
Edna Conway

2 IoT: Privacy, Security, and Your Civil Rights.   15
Cynthia D. Mares

3 Privacy in the New Age of IoT   37
Qi Pan

**4 A Business Framework for Evaluating Trust
in IoT Technology**   53
Fen Zhao and Britt Danneman

**5 Ahead of the Curve: IoT Security, Privacy, and Policy
in Higher Ed**   73
Joanna Lyn Grama and Kim Milford

6 Trust, Identity, Privacy, and Security for a Smart Campus   87
Karen Herrington

7 Security for Science: How One Thing Leads to Another.   97
Hannah Short

8 The Dark Side of Things.   107
Licia Florio

**9 Public Safety and Protection by Design: Opportunities
and Challenges for IoT and Data Science**.   119
Alicia D. Johnson, Meredith M. Lee, and Soody Tronson

10 Privacy Management in the Internet of Things (IoT)   131
Grace Wilson Caudill

11 Securing IoT Data with Pervasive Encryption................... 141
    Eysha Shirrine Powers

12 Secure Distributed Storage for the Internet of Things 159
    Sinjoni Mukhopadhyay

13 Profiles of Women Securing the Future with TIPPSS for IoT 175
    Florence D. Hudson

Index... 185

About the Authors

Alicia D. Johnson is the Resilience and Recovery Manager for the San Francisco Department of Emergency Management. Her work uses human-centered design principles to build collaborative relationships to protect the people and places we value. Alicia led the team that created SF72 and its open source counterpart City72. She has responded to several disasters and large events, regularly serving as an Emergency Operations Center Manager for the City of San Francisco. Alicia holds a bachelor's degree in Communications and Political Science and a master's degree in Public Administration from the University of Colorado. She currently serves as a senior fellow for the West Big Data Innovation Hub.

Britt Danneman is an investor at the Los Angeles-based venture capital firm Alpha Edison. She primarily invests at the Seed to Series B stages to help set business model, fundamental strategy, team and capital plan. She invests thematically, with one such focus on trust. Prior to joining Alpha Edison, Britt worked in distressed and middle market investing at Bain Capital Credit in Boston. She also worked in corporate development and strategy in San Francisco at the fintech startup Funding Circle. She received her MBA from Harvard Business School and undergraduate degree in Finance and Management from The Wharton School at the University of Pennsylvania.

Cynthia D. Mares is a district court judge in the 18th Judicial District of Colorado. She is Senior Legal Advisor and Advisory Board Member for Axon Global, a cybersecurity company located in Houston, Texas. She is also Vice Chair for the Colorado Gaming Commission. Judge Mares is a 2016 alumna of the Harvard Kennedy Executive Education program and a governance fellow with the National Association of Corporate Directors since 2016. She is past president of the Hispanic National Bar Association and Colorado Hispanic Bar Association.

Edna Conway serves as Cisco's Chief Security Officer, Global Value Chain, creating clear strategies to deliver secure operating models for the digital economy. She has built new organizations delivering cyber security, compliance, risk management, sustainability and value chain transformation. She drives a comprehensive security architecture across Cisco's third-party ecosystem. Recognition of her industry leadership includes: membership in the Fortune Most Powerful Women community, a Fed 100 Award, Stevie "Maverick of the Year Award", a Connected World Magazine "Machine to Machine and IOT Trailblazer" Award, an SC Media Reboot Leadership Award, a New Hampshire TechProfessional of the Year Award and CSO of the Year Award at RSA. She holds a JD from the University of Virginia School of Law, and a bachelor's degree from Columbia University, with executive education at Stanford University, MIT and Carnegie Mellon University. Prior to Cisco, she was a partner in an international private legal practice and served as Assistant Attorney General for the State of New Hampshire.

Eysha Shirrine Powers is a Senior Software Engineer at IBM Corporation. She is a Cryptographic Software Developer with 15 years of experience in IBM Z Cryptography and Security. She joined IBM with a Bachelor of Computer Science from the University of Illinois at Urbana-Champaign. After joining IBM, she continued her education with a Master of Information Technology from Rensselaer Polytechnic Institute. Eysha is a prominent speaker for IBM Z Crypto at conferences in the USA and abroad and has several cryptography patents.

Fen Zhao is the Head of Data Science Research at Los Angeles-based venture capital firm Alpha Edison. Prior to joining Alpha Edison, she developed public–private partnerships at the National Science Foundation. She created and led the Big Data Hubs and Spokes Program and was the program coordinator for the Secure and Trustworthy Cyberspace (SaTC) program. During the Obama Administration, Dr. Zhao was an AAAS Fellow at the White House Office of Science and Technology Policy (OSTP). Formerly, Dr. Zhao was an associate with McKinsey's Risk Management Practice. Fen received her PhD in Computational Astrophysics from Stanford University and her BS in Physics from MIT.

Florence D. Hudson is founder and CEO of FDHint, LLC, consulting in advanced technologies and diversity and inclusion. Formerly IBM Vice President and Chief Technology Officer, Internet2 Senior Vice President and Chief Innovation Officer, and an aerospace engineer at Grumman and NASA, she is Special Advisor to the NSF Cybersecurity Center of Excellence at Indiana University, and Northeast Big Data Innovation Hub at Columbia University. She serves on boards for Princeton, Cal Poly, and Stony Brook Universities, and Blockchain in Healthcare. She graduated from Princeton University with a BSE in Mechanical and Aerospace Engineering, with executive education at Harvard Business School and Columbia University.

Grace Wilson Caudill is a United States Agency for International Development (USAID) scholar and an NSF EPSCoR fellow. While attending Kentucky State University, an HBCU in Kentucky, she earned an Associate of Science degree in Electronics Technology, two Bachelor of Science degrees in Computer Science and Network Engineering, and a Master of Science degree in Information Security and Assurance. Grace has embarked on earning her PhD in cybersecurity. Grace currently functions as an IT auditor of cybersecurity at the University of Kentucky. She engages as a consultant with businesses and individuals on Cybersecurity for a Women-owned Small Business - FSS Technologies (FSST). Grace recently held the position of CyberInfrastructure Engineer at the University of New Hampshire. She presented at the national level in academia at various conferences in the USA: conducting Birds of a Feather sessions (BoFs), leading presentations, participating in workshops and hosting discussions around high-performance computing and cybersecurity. Alongside her career and scholarly endeavors, she enjoys hiking, playing volleyball and recreational golf, and volunteers as a YMCA certified swim official for Kentucky and Ohio.

Hannah Short specializes in Trust, Identity and Security for Science. Although based at the European Organisation for Nuclear Research (CERN), on the border of Switzerland and France, she spends most of her time collaborating with a network of colleagues from the Research and Education sector around the globe. After completing a master's degree in Astrophysics, Hannah decided to pursue her newfound interest in programming by becoming a software developer. Since that point, many and varied projects in commercial and research organizations have come her way, leading to her current position at CERN. As more of our lives are spent online, security and privacy have become areas that Hannah prioritizes, for both the technical and ethical challenges. A particular topic of focus is security for distributed authentication systems, for which Hannah received a GEANT Community Award following her contribution to a trust framework for security incident response. In addition, Hannah chairs the Steering Committee for the WISE Community, Wise Information Security for collaborating E-Infrastructures. WISE provides a forum for security representatives from e-Infrastructures to share best practices and, most importantly, to meet face-to-face and build trust between one another.

Joanna Lyn Grama is Senior Consultant with Vantage Technology Consulting Group, where she advises clients on information security policy, compliance, governance, and data privacy issues. She is a member of the US Department of Homeland Security's Data Privacy and Integrity Advisory Committee, appointed by former Secretary of Homeland Security Janet Napolitano, and serves as the chairperson of its technology subcommittee. Joanna received her JD from the University of Illinois College of Law with honors and her BA in International Relations from the University of Minnesota.

Karen Herrington is Director of Information Technology Analytics and Visualization at Virginia Tech. An Information Technology professional with over 30 years of experience, she is a proven leader, having been at the forefront of enabling transformative technologies in both the private sector and the Higher Education arena. Karen's areas of expertise include identity management, Internet of Things, multifactor authentication, data management and analytics. She holds both a Master's and a bachelor's degree in Computer Science from Mississippi State University.

Kim Milford began serving as Executive Director of the Research & Education Networking Information Sharing & Analysis Center (REN-ISAC) at Indiana University (IU) in 2014. She works with members, partners, sponsors, and advisory committees to direct strategic objectives in support of members, providing services and information that allow higher educational institutions to better defend local technical environments, and is responsible for overseeing administration and operations. She led an eMBA course, "Managing Information Risk and Security" for IU's Kelley School of Business. Since joining Indiana University in 2007, Milford has served in several roles leading strategic IT initiatives. As Chief Privacy Officer, she coordinated privacy-related efforts while serving on IU's Assurance Council, chairing the Committee of Data Stewards, and directing the work of the University Information Policy Office including IU's IT incident response team. From 2005 to 2007, Milford worked as Information Security Officer at the University of Rochester leading an information security program that included disaster recovery planning, identity management, incident response and user awareness. In her position as Information Security Manager at University of Wisconsin—Madison from 1998 to 2005, she assisted in establishing the university's information security department and co-led in the development of an annual security conference. Milford provides cybersecurity, information policy, and privacy expertise and presentations at national and regional conferences, seminars and consortia. She has a BS in Accounting from Saint Louis University in St. Louis, Missouri and a JD from John Marshall Law School in Chicago, Illinois.

Licia Florio works for the GÉANT Association as a Senior Trust and Identity Manager. Over the last 15 years, Licia has been involved in many key initiatives that make up the current European and global Authentication and Authorisation Infrastructure for Research & Education (R&E). She supported the Task Force that produced the first eduroam (federated access to wireless networks) pilot—a service that now counts tens of thousands of hotspots in 89 countries across the world; she led the working group that created REFEDS, the global forum that gather R&E identity federations that now counts 90 R&E Identity federations worldwide; she managed the European Funded project on Authentication and Authorisation for Research Collaboration (AARC) to enable federated access for large-scale research collaborations. Currently, she co-leads the Trust and Identity activities in the context of the GEANT project. In June 2018, Licia was awarded the prestigious Medal of Honour by the Vietsch Foundation that supports research and development of advanced Internet technology for scientific research and higher education.

Meredith M. Lee is Founding Executive Director of the West Big Data Innovation Hub. Based at University of California—Berkeley, Dr. Lee develops public–private partnerships and strategic initiatives across industry, academia, nonprofits and government. Under the Obama Administration, she led the White House Innovation for Disaster Response & Recovery Initiative and was an AAAS Fellow at the Homeland Security Advanced Research Projects Agency. She completed her PhD in Electrical Engineering at Stanford University and serves on advisory boards including NASA DIRECT STEM, the Optical Society of America, and the National Leadership Council of the Society for Science and the Public.

Qi Pan is a Digital Media Associate on the Future Leaders Programme at GlaxoSmithKline (GSK), a world-leading healthcare company. She was the GDPR expert for Consumer Healthcare Tech, responsible for training and successfully rolling out GDPR-compliant technologies to the GSK Consumer Healthcare salesforce across EU markets in 2018. Qi organizes thought-leadership debates in GSK, where external experts are invited to bring the outside in and inspire transformative ways to benefit consumers and patients. Prior to GSK, Qi studied Molecular and Cellular Biochemistry at the University of Oxford, graduating with an MBiochem for her research into the role of epigenetics in X chromosome inactivation. Aside from work, Qi is a staunch advocate of inclusion and diversity with a focus on women in STEM.

Sinjoni Mukhopadhyay is a fourth year Computer Science PhD candidate at the University of California, Santa Cruz. Prior to beginning her PhD career, she completed her Bachelor's in Electronics and Telecommunication from India, followed by her Master's (Efficient Reconstruction Techniques for Disaster Recovery in Secret-Split Datastores) in Computer Science from the University of California, Santa Cruz. Sinjoni's areas of interest include storage security, archival storage, distributed and cloud storage. She is currently working on building a self-improving synthetic workload generator using neural networks that can generate workloads which can be used to test predicted future systems. In her free time she usually reads novels, paints or practices Indian classical dance forms like Bharatnatyam and Odissi.

Soody Tronson has over 25 years of operational experience in technology, business, management, and law in start-up and fortune 100 companies. She is founding managing counsel at the law firm of STLGip, and founder and CEO of Presque, a wearable consumer medical device company. She serves in board and advisory capacities with STEM to Market national accelerator, California Lawyers Association Executive Committee of the Intellectual Property Section, and Licensing Executives Society USA/Canada Women in Licensing Committee. Soody holds a JD, MS in industrial chemistry, and BS in chemistry, and is licensed to practice before the State of California and the US Patent and Trademark Office. In her role as a licensing executive, she is leading efforts to provide best practices and recommendations on streamlining data sharing arrangements.

Abbreviations

AAAS	American Association for the Advancement of Science
AARC	Authentication and Authorisation for Research Collaboration
ABAC	Attribute-Based Access Control
ACE	Authorization for Constrained Environments
ACV	Automated Cryptographic Validation
ADMQP	Advanced Message Queuing Protocol
AES	Advanced Encryption Standard
AES-CBC	Advanced Encryption Standard - Cipher Block Chaining
AI	Artificial Intelligence
AMQP	*Advanced Message Queuing Protocol*
API	Application Programming Interface
ASCII	American Standard Code for Information Interchange
ASU	Arizona State University
AUP	Acceptable Use Policy
AWS	Amazon Web Services
BA/AB	Bachelor of Arts degree
BOF	Birds of a Feather session
BS	Bachelor of Science degree
BSE	Bachelor of Science in Engineering degree
BYOD	Bring Your Own Device
BYOE	Bring Your Own Everything
CAGR	Compound Annual Growth Rate
CalECPA	Electronic Communications Privacy Act of California
CBC	Cipher Block Chaining
CCPA	California Consumer Privacy Act of 2018
CEO	Chief Executive Officer
CERN	European Council for Nuclear Research
CF	Coupling Facility
CFRM	Coupling Facility Resource Management
CIA	US Central Intelligence Agency
CIA Triad	Confidentiality, Integrity and Availability

CIPT	Certified Information Privacy Technologist
CISA	Cybersecurity Information Sharing Act of 2015
CISSP	*Certified Information Systems Security Professional*
CLOUD Act	Clarifying Lawful Overseas Use of Data Act
CMR	Centralized Multi-node Repair
CNN	Cable News Network
CoAP	Constrained Application Protocol
CPACF	Central Processor Assist for Cryptographic Function
CPR	Cardio Pulmonary Resuscitation
CPU	Central Processing Unit
CRC	Campus Recreation Complex
CRISC	Certified in Risk and Information Systems Control
CRT	Chinese Remainder Theorem
CSO	Chief Security Officer
CTI	Cyber Threat Indicators
DBA	Data Base Administrator
DDoS	Distributed Denial of Service
DES	Data Encryption Standard
DM	Defensive Measures
DR	Disaster Recovery
DRBG	Deterministic Random Bit Generator
DSRC	Dedicated Short Range Communications
DTSA	Defend Trade Secrets Act of 2016
EAP-SIM	Extensible Authentication Protocol - Subscriber Identity Module
EBCDIC	Extended Binary Coded Decimal Interchange Code
ECB	Electronic Code Book
ECC	Elliptic Curve Cryptography
EEA	European Economic Area
EFF	Electronic Frontier Foundation
eMBA	electronic Master of Business Administration
ENISA	European Network and Information Security Agency
EPSCoR	Established Program to Stimulate Competitive Research
ETC	Electronic Toll Collection
EU	European Union
FBI	Federal Bureau of Investigation
FIAM	Federated Identity and Access Management
FIM	Federated Identity Management
FIPPs	Fair Information Practice Principles
FIPS 140-2	Federal Information Processing Standard Publication 140-2
FPE	Format Preserving Encryption
FTC	Federal Trade Commission
GDPR	General Data Protection Regulation
GE	General Electric
GEANT	European Union Research and Education Network
GPS	Global Positioning System

GPU	Graphics Processing Unit
GRC	governance, risk, and compliance
GSK	GlaxoSmithKline
HBCU	Historically Black Colleges and Universities
HCI	Human-Computer Interaction
HIPAA	Health Information Portability and Accountability Act of 1996
HITECH	Health Information Technology for Economic and Clinical Health Act
HMO	Health Maintenance Organization
HP	Hewlett Packard
HR	Human Resources
HSM	Hardware Security Module
HTTP	Hyper Text Transfer Protocol
HW	Hardware
IAM	Identity and Access Management
ICT	Information and Communications Technology
ID	Identification or Identifier
IdP	Identity Provider
IEEE	Institute of Electrical and Electronics Engineers
IETF	Internet Engineering Task Force
IMEI	International Mobile Equipment Identity
IaaS	Infrastructure as a Service
IoT	Internet of Things
IP	Intellectual Property
IP Address	Internet Protocol Address
IPv6	Internet Protocol version 6
IPR	Intellectual Property Rights
ISAC	Information Sharing and Analysis Center
ISAO	Information Sharing and Analysis Organization
ISO 19790	International Organization for Standardization, Security requirements for cryptographic modules
IT	Information Technology
ITU-T	International Telecommunication Union (ITU) Telecommunication Standardization Sector (T)
IU	Indiana University
JCE	Java Cryptography Extension
JD	Juris Doctor of Law degree
JVM	Java Virtual Machine
LAN	Local Area Networking
LHC	Large Hadron Collider
LiDAR	Light Detection and Ranging
LIGO	Laser Interferometer Gravitational wave Observatory
MAC	Media Access Control
MBA	Master of Business Administration degree
MFA	Multi Factor Authentication

MGI	McKinsey Global Institute
MIT	Massachusetts Institute of Technology
MI5	UK Security Service
MK	Master Key
ML	Machine Learning
MQTT	Message Queuing Telemetry Transport
MRC	Medical Research Council in the UK
MS	Master of Science degree
MSIS	Multi-Secret Image Sharing
M2M	Machine-to-Machine
NACD	National Association of Corporate Directors
NFC	Near Field Communications
NIST	National Institute of Standards and Technology
NIST/CSE	National Institute of Standards and Technology (NIST)/ Communications Security Establishment (CSE)
NLP	Natural Language Processing
NOC	Network Operating Center or Network Operations Center
NP	Non-deterministic Polynomial-time
NSA	National Security Agency
OASIS	Organization for the Advancement of Structured Information Standards
OAuth	Open Authentication Protocol
ODM	Original Device Manufacturer
OEM	Original Equipment Manufacturer
OECD	Organisation for Economic Co-operation and Development
OIDC	OpenID Connect
OSTP	Office of Science and Technology Policy
OT	Operational Technology
PaaS	Platform as a Service
PCI-DSS	Payment Card Industry Data Security Standard
PCIe	Peripheral Component Interconnect Express
PhD	Doctor of Philosophy degree
PHI	Protected Health Information
PII	Personal/Personally Identifiable Information
PKC	Public Key Cryptography
PKI	Public Key Infrastructure
PUF	Physically Unclonable Function
P2P	Peer to Peer
R&E	Research & Education
REFEDS	Research and Education FEDerations group
REN-ISAC	Research and Education Networking Information Sharing and Analysis Center
REST	Representational State Transfer
RF COMMS	Radio Frequency Communications
RFID	Radio Frequency Identification

RSA	Rivest–Shamir–Adleman public-key encryption technology
SAML	Security Assertion Markup Language
SaTC	Secure and Trustworthy Cyberspace
SCA	Stored Communications Act
SDL	Secure Development Lifecycle
SEC	Securities and Exchange Commission
SKA	Square Kilometer Array
SaaS	Software as a Service
SIM	Subscriber Identification Module
SOC	Security Operating Center or Security Operations Center
SP	Service Provider
SQL	Structured Query Language
SSO	Single Sign On
STEM	Science, Technology, Engineering and Mathematics
SURF	Collaborative ICT organisation for Dutch education and research
Tbps	Terabits per second
TCO	Total Cost of Ownership
TCP	Transmission Control Protocol
TIPPSS	Trust, Identity, Privacy, Protection, Safety, Security
TNC2018	Terena Networking Conference 2018 in Europe
TOR	The Onion Router
TPM	Trusted Platform Module
TTPS	Techniques, tactics, and practices
TV	TeleVision
UDP	User Datagram Protocol
UK	United Kingdom
ULC	Uniform Law Commission
UMA	User-Managed Access
UN	United Nations
US and USA	United States of America
USAID	United States Agency for International Development
UTSA	Uniform Trade Secrets Act
VPN	Virtual Private Network
VSD	Value Sensitive Design
V2V	vehicle-to-vehicle communication
WiDS	Women in Data Science
WISE	Wise Information Security for collaborating E-Infrastructures
WSN	Wireless Sensor Networks
XACML	eXtensible Access Control Markup Language
XML	eXtensible Markup Language
XMPP	eXtensible Messaging and Presence Protocol
X.509	international public key infrastructure (PKI) standard
YMCA	Young Men's Christian Association
6LoWPAN	IPv6 over Low-power Wireless Personal Area Networks
802.1X	IEEE Standard for port-based Network Access Control

Chapter 1
IoT: Is It a Digital Highway to Security Attacks?

Edna Conway

1.1 Introduction to a Methodology to Secure the IoT Ecosystem

It used to be that we would imagine a world where things talk to, listen to, and observe all of us, so we could better understand ourselves and others, where biometric data about us might be compiled in real time as we eat, sleep, and go about our lives in order to provide us with better health outcomes—a world where devices could talk to other devices at speeds beyond human comprehension to improve the performance of auto and air travel, factory floor production, or even just the email on our phones. With today's Internet of Things (IoT) we are already roaring down that very digital highway! Our challenge is how to reap the benefits of that connected world while also ensuring security with every IoT connection we make.

For purposes of this discussion, let us agree that IoT, at its core, is what the Institute of Electrical and Electronics Engineers (IEEE) concluded in 2015 for low complexity systems. IoT is "a network that connects uniquely identifiable 'Things' to the Internet. The 'Things' have sensing/actuation and potential programmability capabilities. Through the exploitation of unique identification and sensing, information about the 'Thing' can be collected and the state of the 'Thing' can be changed from anywhere, anytime, by anything [1]."

200 billion [2]. 200 billion is the number of devices that are predicted to be digitally connected by 2020, that is, more than 22 devices for every one of us who will be on planet earth by then. Who and what are making, operating, and accessing these connected devices?

E. Conway (✉)
Merrimack, NH, USA
e-mail: ednaconway@gmail.com

© Springer Nature Switzerland AG 2019
F. D. Hudson (ed.), *Women Securing the Future with TIPPSS for IoT*, Women in Engineering and Science, https://doi.org/10.1007/978-3-030-15705-0_1

Table 1.1 IoT application areas

IoT application area	Description
Connected vehicles	IoT enabling vehicles and transportation infrastructure (e.g., roadway, traffic lights, cameras) to communicate.
Consumer IoT	IoT in the home and wearable and mobile connected devices.
Health IoT	IoT which processes data derived from sources such as electronic health records and patient generated health data.
Smart buildings	IoT such as energy usage monitoring systems, physical access control security systems, and lighting/temperature control systems.
Connected factories	IoT integrating real-time operations data, facilitating equipment function and monitoring, quality control, and failure analysis.

What will these 200 billion connected devices be doing? They will be sharing information and controlling operations across a spectrum we could not have imagined even 5 years ago. This convergence of Information Technology (IT) and Operational Technology (OT) has been sweeping global industries, including sectors such as energy, heavy equipment, and transportation. IoT has also expanded into all aspects of daily living and government, exacerbating the need for ever more vigilance and security across and through the IoT environment.

Before outlining a methodology to drive security across the IoT environment, it is helpful to categorize a few key IoT technology application areas. The United States National Institute of Standards & Technology (NIST) has identified five such areas in its Draft Interagency Report 8200 [3], which seeks to identify and list the many international cybersecurity standards that are applicable to IoT. Application Area descriptions follow in Table 1.1.

While we look to the future promise of exponential IoT growth, we must be prepared for the corollary security challenge. Of the utmost concern is this hidden and often overlooked reality: *as we digitize we are expanding the ecosystem of third parties who will inevitably impact us*, who will be "touching our stuff" along the Internet highway. For better, or for worse, the more we connect—the more transparent and collaborative we are—the more we are allowing others to observe and possibly control us.

As participants in digital transformation, whether individually or at an enterprise level, we must be aware of who and what is digitally and physically touching our information and devices. I call this the "third-party ecosystem[1]."

An interpretation of *Ponemon Institute's March 2018 Second Annual Study on IoT* [4] revealed a glaring reality regarding the security risk from the exponential growth of devices provided by the third-party ecosystem. The risk of the unknown is prevalent. Respondents can only fully identify less than 10% of devices connected to their networks. What is unknown cannot be secured (Fig. 1.1).

---

[1] Throughout this chapter, all references to the third-party ecosystem or ecosystem, by definition, include a community of third parties who are part of the Internet of Things (IoT).

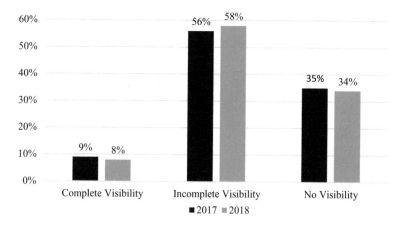

Fig. 1.1 Data answering the question "Are you aware of the network of physical objects that are connected?" Ponemon Institute's Second Annual Study on IoT, March 2018 [4]

Successful navigation of the digital super highway requires three key steps:

- Understand the security threats and their impact.
- Identify who is doing what within the connected ecosystem.
- Deploy a set of pervasive security techniques and processes across that ecosystem.

1.2 Threats and Related Exposures in the Connected Ecosystem

The connected ecosystem is increasingly the source of attacks. Disruption and disclosure of confidential information by third parties with whom we are connected, knowingly or unknowingly, continues to expand, as shown in Fig. 1.2.

Beyond reported attacks, research across global enterprises offers richer insight into the third-party impact. All third-party impact is significant. The data demands a call to action: 75% of the time incidents can be attributed to third parties (Fig. 1.3).

Third-party IoT devices are expanding overall third party risk. As the deployment of IoT devices expands, the related third-party security risk of data loss and cyberattacks from those devices will only rise. While some certainty of causation exists today, data shows a level of uncertainty that will increase the risk of unprotected IoT devices (Fig. 1.4).

Clearly the third-party ecosystem security risk is poised to grow. Sixty percent of respondents to the *Ponemon Institute's Second Annual Study on the IoT*, indicated their enterprises have a third-party risk management program. Forty-two percent of these respondents said the program is part of their companies' enterprise risk management program, but only 29% of respondents said their enterprises actively monitor the risk of IoT devices used by third parties [4].

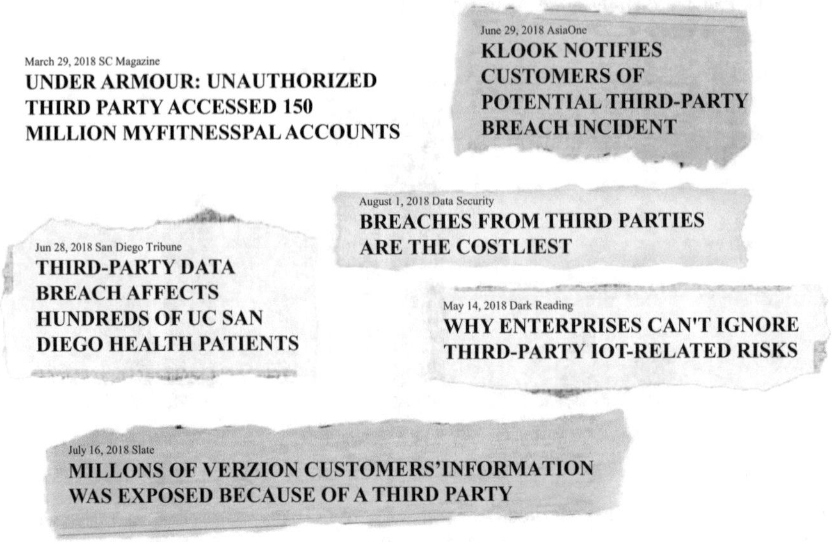

March 29, 2018 SC Magazine
UNDER ARMOUR: UNAUTHORIZED THIRD PARTY ACCESSED 150 MILLION MYFITNESSPAL ACCOUNTS
June 29, 2018 AsiaOne
KLOOK NOTIFIES CUSTOMERS OF POTENTIAL THIRD-PARTY BREACH INCIDENT
August 1, 2018 Data Security
BREACHES FROM THIRD PARTIES ARE THE COSTLIEST
Jun 28, 2018 San Diego Tribune
THIRD-PARTY DATA BREACH AFFECTS HUNDREDS OF UC SAN DIEGO HEALTH PATIENTS
May 14, 2018 Dark Reading
WHY ENTERPRISES CAN'T IGNORE THIRD-PARTY IOT-RELATED RISKS
July 16, 2018 Slate
MILLONS OF VERZION CUSTOMERS'INFORMATION WAS EXPOSED BECAUSE OF A THIRD PARTY

Fig. 1.2 News abounds with risks linked to third parties

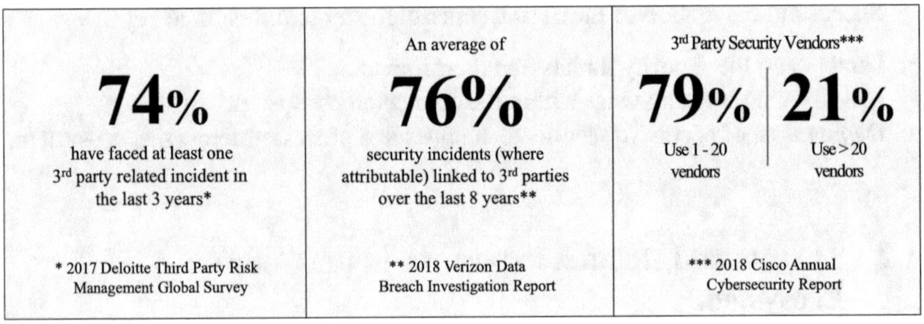

74%
have faced at least one 3rd party related incident in the last 3 years*
* 2017 Deloitte Third Party Risk Management Global Survey
An average of
76%
security incidents (where attributable) linked to 3rd parties over the last 8 years**
** 2018 Verizon Data Breach Investigation Report
3rd Party Security Vendors***
79%
Use 1 - 20 vendors
21%
Use > 20 vendors
*** 2018 Cisco Annual Cybersecurity Report

Fig. 1.3 Third parties—a critical source of security risk

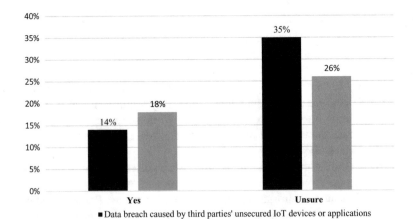

40%
35%
30%
25%
20%
15%
10%
5%
0%
14%
18%
35%
26%
Yes
Unsure
■ Data breach caused by third parties' unsecured IoT devices or applications

Fig. 1.4 Data answering the question "Has your company experienced a data breach or cyber attack caused by a third party's unsecured IoT devices?" Ponemon Institute LLC Second Annual Study on IoT, Publication Date: March 2018 [4]

To meaningfully address this inevitability, let us step back and examine the problem by defining it in terms of threats and threat impacts.

1.2.1   The Threats

Manipulation—The alteration of technology that allows unintended control or observation. Such an alteration of an IoT device and its resultant security vulnerabilities can have a host of ramifications. Ramifications that include a failure of the IoT device itself or control of the Information Technology (IT) systems to which it connects, including a denial of service. Ramifications can also manifest in the Operational Technology (OT) that has converged with these affected IT systems, including outright failures or reconfigured operational settings.

Espionage—The observation of confidential information at any point in the new ecosystem of digitally and operationally converged technology. Espionage is not just the prerogative of nation states anymore.

Disruption—Whether the most draconian level of a full denial of service or precise surgical alterations that allow data and operational processes to be changed.

1.2.2   The Threat Impacts

Tainted Solutions—Whether hardware, software or cloud-based services, the threats identified above lead to the risk of taint. Something that no longer functions as its designer or user intended. Taint can have far-reaching consequences.

Counterfeit Solutions—Functional integrity and quality are compromised when deceptively "real" looking and functioning technology is put into operation.

Intellectual Property Misuse—The lifeblood of innovation, intellectual property (IP), when disclosed in whole or in part, can be effectively leveraged by bad actors to manipulate, falsify, and create tainted and counterfeit solutions.

1.3   Understand Who and What Comprises the Third-Party Ecosystem

Having identified the threats and exposures, the next step to successfully navigating the connected ecosystem is to (1) identify the key players in your third-party ecosystem and (2) understand what those third parties deliver to you.

The Information and Communications Technology (ICT) third-party ecosystem is core to the digital convergence of IT and OT. Moreover, it serves as an illustrative example of both the sheer vastness and diversity of that ecosystem. Members of that ICT third-party ecosystem are depicted in Fig. 1.5 below.

Open Source Software	Software Licensors	HW Component Suppliers	Cloud Service Providers
Logistics Partners	OEMs/ODMs	IoT Devices	Manufacturing Partners
Channel/Distribution	Repair/Refurbishment Partners	Scrap Partners	Recycling Partners

Fig. 1.5 Members of the ICT third-party ecosystem

1.3.1 Drive Pervasive Security Across the Third-Party Ecosystem

The diversity of the third parties who participate in the life cycle of ICT solutions makes one thing clear. Pervasive security, namely the right security in the right way at the right time, can only be achieved if we coordinate meaningfully with those third parties.

To achieve the necessary level of coordination, we must develop a common taxonomy, as we did with the three threats and impacts mentioned earlier. Beyond that, a flexible architecture that can effectively be deployed across and through this diverse third-party ecosystem is essential.

1.3.2 A Flexible Security Architecture

A good approach is to establish key security architecture domains that can be deployed across the ICT third-party ecosystem. Most importantly, these domains should be agreed upon by all and be flexible enough to be adapted to fit the needs of all ICT third parties.

Brief descriptions and examples of Core Domains are listed in Table 1.2.

Leveraging an architecture touching upon these domains can allow third parties to effectively collaborate and drive comprehensive security. The domains can also serve as an approach to embedding security (including cybersecurity) into procurement [5].

It cannot be said too often: Security is a Team Sport. While the overarching architecture addresses all third parties, it must be flexible enough to allow variability. This variability allows for customized goals based on the nature of the products or services received from each specific third party (e.g., printed circuit board Gerber files or integrated circuit masks).

A key to success is to establish flexible security goals within each relevant domain, rather than setting forth specific requirements. In other words, keep security non-prescriptive to the optimum extent possible. Only by collaborating to understand the rich variety of third-party business models can we enable security

Table 1.2 Examples of core domains and descriptions

	Domain	Description
1	Security Governance	The security governance domain details requirements for an overall governance strategy to manage value chain security and compliance related risks by establishing requisite policies, standards, and procedures.
2	Security in Manufacturing and Operations	The security in manufacturing and operations domain details requirements for manufacturing and operating procedures in order to protect material assets, intellectual property, and information.
3	Asset Management	The asset management domain details requirements for securing IT and manufacturing assets throughout their life cycle.
4	Security Incident Management	The security incident management domain details requirements to establish a robust incident management process that should be followed for activities such as logging, recording, and resolving security incidents and anomalies.
5	Security Service Management	The service management domain details requirements: (a) for the delivery of services in accordance with agreed upon delivery timeframes, quality and security levels (b) for establishing a business continuity plan/program in the event of service disruption
6	Security in Logistics and Storage	The security in logistics and storage domain details security requirements that should be followed during storage and distribution of raw materials, inventory, and finished goods.
7	Physical and Environmental Security	The physical and environmental security domain details requirements that value chain members must design and implement to control access to facilities, equipment and resources, and to protect personnel and property from damage, harm, or unauthorized alteration.
8	Personnel Security	The personnel security domain details requirements to ensure that all value chain personnel who have access to any proprietary items, intellectual property and confidential information have the required authorizations, training, and contractual agreements including appropriate clearances, if required.
9	Information Protection	The information protection domain details requirements for protection of proprietary data through its life cycle, such as data classification, handling, cryptographic controls, and disposal. It also lists the requirements to be implemented on information systems that store or process intellectual property.
10	Security Engineering and Architecture	The security engineering and architecture domain details requirements to be followed during design, development, testing, and rollout of products (tangible and intangible) and services.
11	3rd Tier Partner Security	The third-tier partner security domain details requirements focused on information security controls that must be implemented at downstream value chain members (fourth parties, e.g., cloud service providers) in relation to procurement of goods and services.

that is embedded in the tools, processes, and people of the ecosystem. Flexible goals enable greater third-party adoption and swifter deployment.

For example, a prescriptive requirement addressing passwords might look like this:

Supplier must implement access controls on Information Systems via strong passwords and unique individual identifiers that are not shared among multiple users. Passwords must contain:

- At least eight alphanumeric characters;
- Both upper and lower case letters;
- At least one number (e.g., 0–9); and
- At least one special character (e.g., !$%^&*()_+|~-=\`{}[]:";'<>?,/).

Further, the following practices must also be adhered to, at a minimum:

- Passwords must be changed at least every 180 days.
- After five failed login attempts a system alert must be created.
- Information Systems must prevent the reuse of the last ten passwords.
- Passwords must not be shared.

Alternatively, a flexible goal-based approach might state that access control must be implemented via a combination of multifactor authentication techniques. Such authentication can be any of the following:

- Biometric, mobility, or human behavioral based (e.g., fingerprint or swiping motion) and
- Incorporate traditional strong alphanumeric-character passwords of unlimited length or passphrases without duplicated words (aka "memorized secret message" according to NIST [6]) or randomly generated passwords.

Let's explore exactly how such an architecture might work with regard to IoT. To do that we can use cryptography as a discussion point around Domain 9, Information Protection, as defined in Table 1.2. Examples of Core Domains and Descriptions.

1.3.2.1   A Cryptography Example of Domain 9 (Information Protection)

In the truly digitized environment that we are racing toward, encryption is a building block of security. We are aware of the risks of intentionally altered or improperly implemented encryption. Public–private effects have focused on validating the accuracy of the algorithm implementing the encryption.

NIST spearheaded a program designed to address validation of cryptographic modules (Fig. 1.6).

Applying the concept of driving the right security at the right time requires an understanding of what the "right way" might look like. Leveraging and validating encryption in the IoT environment requires us to think through the unique function and operational parameters of the device itself, where it is located and what its purpose is. Of unique value is applying open protocols for Automated Validation of Encryption (Fig. 1.7).

Implementing encryption in an IoT environment in the same manner encryption is implemented in large capacity compute environments is a ticket to failure.

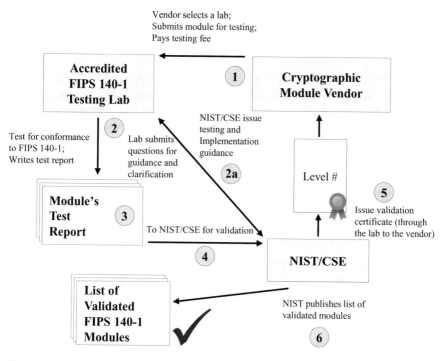

Fig. 1.6 Cryptographic Module Validation Process (Courtesy of NIST [7])

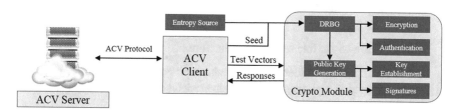

Fig. 1.7 Implementing encryption in an IoT environment. *ACV* = Automated Cryptographic Validation, Courtesy of Cisco Systems, Inc. [9]

Understanding the third-party device and its limitations and constraints is essential. Applying an Automated Cryptographic Validation Protocol can enhance efficiency and secure operation. However, we must evaluate the unique environment.

Recognizing the benefits of encryption in constrained environments such as automotive systems, sensor networks, healthcare, distributed control systems, the Internet of Things (IoT), cyber-physical systems, and the smart grid, NIST put forth an informational report on Lightweight Cryptography [8].

"Constrained environments cannot always use all the commonly accepted crypto algorithms available because of their constrained nature. A battery operated sensor, for example, cannot use 3072-bit RSA because it would deplete its battery faster and because of the processing load [9]."

Introducing Advanced Cryptographic Validation in Lightweight Encryption for constrained environments, such as IoT, is a glaring example of a flexible security architectural approach. An example of such an Advanced Cryptographic Validation Protocol that operates for lightweight crypto can be found at https://github.com/sigmaJ/ncsu-wolfssl. Applying security practices, modified for the IoT environment, delivers higher integrity overall while retaining the operational efficiency of IoT devices.

This kind of architectural approach can serve to further enhance the security posture of IoT.

1.4   Deploy the Security Architecture Using a Layered Approach

Thus far on the journey the following steps have been addressed:

- Establishing a taxonomy of threats and related exposures in the connected ecosystem.
- Understanding who and what comprises the third-party ecosystem—using the ICT ecosystem as an example.
- Developing a process to drive pervasive security across the third-party ecosystem:

  - Establishing a flexible security architecture.

A flexible architecture alone is not enough. Deployment of the architecture using a layered approach is highly recommend. Layering techniques in each of the following areas should be considered:

Physical Security: Deployed from components-to-finished product. Examples include: traceability, real-time transport tracking, security checkpoints, biometric access gates, segregation of high-value materials, tamper resistant labeling and packaging, and role-based access control to all physical locations.

Logical (Operational) Security: Implement rules-based access and leave no device unprotected, from security cameras to personal phones. Examples include: requiring all product development to follow strict Secure Development Lifecycle (SDL) protocol, encrypt data transmissions, conduct material reconciliation, and carefully manage all data destruction and scrap handling processes.

Security Technology: Utilize applicable security technology through the stages of the IoT life cycle. Examples include: deploying next generation encryption, anti-counterfeiting chips, insertion of immutable identity during test, secure device boot and deploying obfuscation techniques at the integrated circuit level.

Behavioral Security: Embrace pervasive security cultural behaviors. For example, raise and maintain awareness around phishing campaigns, encourage employee participation in "see something say something" programs, adhere to "carrot vs. stick" management.

Network Security: Approach IoT as a part of the IT network, even if simply an OT sensing device. Examples include: network segmentation for IoT information input, controlled IoT device linking, encrypting IoT device transmission, and vetting security of clouds storing or transmitting IoT device data.

1.5   A Coordinated Deployment Plan

To make driving pervasive security a reality, an enterprise-wide coordinated deployment plan is key. The third-party ecosystem is commercially managed from multiple functions within an enterprise. Imagine, for example, the risk to successful deployment of your architecture without the governance risk and controls/compliance organization or the development of quality teams. Engage EVERYONE across your enterprise, whether your enterprise is commercial, educational or governmental.

A coordinated plan can include the following:

- Building compliance to security architecture into performance management, that is, scorecards and metrics for third parties.
- Active sharing of security best practices and information in public–private partnerships.
- Serving as a liaison to governmental agencies writing or enforcing laws and regulations as they address the challenge of pervasive security across a vast third-party ecosystem.
- Incorporating security parameters into the development life cycle, operational tools, and manufacturing processes.
- Developing processes to effectively evaluate the security maturity of third parties into the onboarding and procurement process for your enterprise.
- Publishing internally all enterprise functions success/failure in ensuring that the third parties they manage are adhering to the security architecture.

After all, security is a highly collaborative team effort and measuring the collective enterprise as a whole can afford more meaningful visibility and security integrity.

1.6   Conclusion: Safely Merging onto the IoT Super Highway

The IoT Super Highway is the path to our hyperconnected world. To ensure our security in this environment, we must keep in mind that at the foundation of IoT lies the network. After all, it is the "Internet" of things. Securing IoT devices and their foundation—the network itself—will ensure true digital transformation.

Application of IoT has the infinite potential to transform business, society and the global economy. We can achieve that potential only by also delivering security at every step along the digital journey.

Leveraging the pitstops and charging stations along the IoT super highway, as outlined in this chapter, will enable pervasive security. Perhaps most importantly, it must be remembered that the steps, summarized below, can only be successful if taken together across the third-party ecosystem:

1. Establish the common set of security threats.
2. Rally around clear goals that can only be achieved collectively.
3. Understand the unique business aspects of key third parties.
4. Openly, and without retribution, reveal challenges.
5. Share technical security strategies, practices and successes.

The extraordinary opportunity to reap life-altering benefits from the burgeoning growth of IoT is ours as we roar down today's digital highway.

References

1. Minerva R, Biru A, Rotondi D (2015) Towards a definition of the Internet of Things (IoT) Revision1. The Institute of Electrical and Electronics Engineers (IEEE), Piscataway, NJ
2. Intel Corporation [Internet]. Santa Clara: The Corporation; c.2014 [cited 2018 Aug 3]. "Intel IoT Gateway." Available from: https://www.intel.com/content/dam/www/public/us/en/documents/product-briefs/gateway-solutions-iot-brief.pdf
3. National Institute of Standards and Technology (NIST) (2018) Interagency report on status of international cybersecurity standardization for the Internet of Things (IoT). National Institute of Standards and Technology (NIST), Gaithersburg, MD
4. Ponemon Institute and The Santa Fe Group (2018) Second annual study on The Internet of Things (IoT): a new era of third-party risk. Ponemon Institute and The Santa Fe Group, Traverse City, MI. Sponsored by the Shared Assessments Program
5. See NIST 2015 Case Study for more information on the Cisco Value Chain Security Architecture: National Institute of Standards and Technology (NIST) (2015) Best Practices in Cyber Supply Chain Management. Cisco: Managing Supply Chain Risks End to End. National Institute of Standards and Technology (NIST), Gaithersburg, MD
6. National Institute of Standards and Technology (NIST) (2017) Special Publication 800-63, Revision 3: Digital Identity Guidelines. National Institute of Standards and Technology (NIST), Gaithersburg, MD
7. National Institute of Standards and Technology (NIST) (2014) NIST ITL Bulletin for November 2014: Cryptographic Module Validation Program (CMVP). National Institute of Standards and Technology (NIST), Gaithersburg, MD

8. National Institute of Standards and Technology (NIST) (2018) Internal Report 8114: Report on Lightweight Cryptography. National Institute of Standards and Technology (NIST), Gaithersburg, MD
9. Kampanakis P (2017) Collaborating with NCSU to promote lightweight crypto validation and assessment. [cited 2018 Aug 3]. In: Cisco Blogs [Internet]. Cisco Systems, Inc., San Jose, CA. Available from: https://blogs.cisco.com/security/collaborating-with-ncsu-to-promote-lightweight-crypto-validation-and-assessment

Chapter 2
IoT: Privacy, Security, and Your Civil Rights

Cynthia D. Mares

2.1 Where Does the Protection of Our Civil Rights Currently Begin, and Where Should It Begin?

In the USA, our rights are protected by the US Constitution and state constitutions. We begin this chapter with the goal of having a basic understanding of how our constitutional rights are affected by the IoT. As has been stated elsewhere in this book, the IoT is a network of physical devices (electronics, vehicles, home appliances, etc.) that are connected via the Internet that can collect and share data across a network. A quick Internet search will tell you that by the year 2020 there will be somewhere around 31 billion devices or more connected to the IoT worldwide. My guess is that it will be even higher than that. Nevertheless, that is a phenomenal number that should catch your attention, especially as an engineer or product developer. Consider this: "Today we are collecting data at an unprecedented rate. The volume, velocity, and variety of data being gathered through the internet and other technologies is estimated to be over 2.5 quintillion bytes of data a day—that's 2.5 followed by a staggering 18 zeros!" [1].

IoT privacy and security concerns are a result of these amazing and creative modes of immense data collection beyond our imagination. We all have devices connected to the IoT, such as smart watches, security systems, heart monitors, in-home IoT early dementia detection apps, etc., which are all items that are now the channel for data collection we thought previously imperceptible. The Economist has declared the world's most valuable resource as no longer oil but data. This declaration is an indication of how valuable your personal information is—just imagine what hackers will do to get it. Protecting your personal data is critical. Artificial

C. D. Mares (✉)
Centennial, CO, USA
e-mail: cdiannemares@gmail.com

© Springer Nature Switzerland AG 2019
F. D. Hudson (ed.), *Women Securing the Future with TIPPSS for IoT*, Women in Engineering and Science, https://doi.org/10.1007/978-3-030-15705-0_2

Intelligence (AI) techniques such as machine learning extract even more value from data, using algorithms which can predict when a customer is ready to buy, a jet-engine needs servicing or a person is at risk of a disease. Industrial giants such as GE and Siemens are now selling themselves as data firms [2].

Your personal data is extremely valuable to companies, even when obtained legally, but once it is stolen, it can then be sold on the Dark Web to be taken advantage of again. This is a place on the Internet where criminals buy and sell your private information and other things, including intellectual property, stolen credit cards, Social Security numbers, medical information, firearms, any drug imaginable, etc. Your healthcare records are on the top of the list. Data from the healthcare industry, which includes both personal identities and medical histories, can live a lifetime. One in 13 patients will have their health records stolen after a healthcare provider data breach, including personal information such as social security or financial records. Cyberattacks are projected to cost US health systems $305 billion in cumulative lifetime revenue from 2015 through 2019, according to a report from Accenture in 2015 [3].

The Dark Web is only accessible through a specific software called TOR, accessible to anyone who wishes to download the software on his or her computer. TOR stands for The Onion Router, which was developed by the US Navy in the 1990s to allow intelligence agents operating overseas to communicate anonymously with colleagues in the USA. Cybersecurity experts can access the dark web safely and even tell you if your personal information is available there. I bring these facts to your attention as a segue into the next issue we will address, which is the world of cybersecurity and litigation caused as a result of the huge expansion of the IoT.

Intellectual property is also very sought after by hackers. In a March 23, 2018 press release, the US Department of Justice reported that Mabna Institute hackers penetrated systems belonging to hundreds of universities, companies, and other victims to steal research, academic and proprietary data, and Intellectual Property [4].

The issue of privacy, or some might say, the lack thereof, is being litigated across the country and worldwide every day. Even more unfortunate are the increasing number of breaches and hacks that occur every single day. Hackers are paid ransom money daily. Consequently, huge amounts of dollars are spent by the government to track down the cyberhackers. As can easily be found on the Internet, the most recent data breaches in 2018 include Under Armour/MyFitness Pal, Orbitz, Panera Bread, and Saks Fifth Avenue/Lord & Taylor.

As an engineer, you are the inventor, the product developer, the creative mind. One of the goals of this chapter is to give you solid reasons why you should not feel blunted if you were forced to include an entire team, including lawyers, front line employees and marketing staff during the innovative process. Would you feel hesitant to include an entire team because you feel like this "team" does not belong in the product development stage?

Take into consideration the fact that new laws are being enacted worldwide every day, many with which you must comply, that require IoT device makers to follow minimum security standards, such as the ability to install remote security updates. Take for example, the European Union's Cybersecurity Act and its General Data

Protection Regulation (GDPR) or the US Senate bill known as the Internet of Things Cybersecurity Improvement Act and the Cybersecurity Act of 2015. Whether US or foreign laws, they all require compliance, and noncompliance can be very costly. Therefore, there is good, solid reason why it may make sense to have a "team" involved at the beginning stages of product development. As an example, if your product does not have the capability to comply with laws that require products to be built with the capability to install remote security updates, you may find yourself out of business. Having the capability to remotely install security updates would be essential to the livelihood of any product and, likely, the company itself. Laws change frequently and are always behind, trying to keep up with technology; therefore, a knowledgeable team at the early stages of product development is critical.

2.2 Privacy and Security Defined

What is privacy and why does it matter? I like this succinct definition: "It relates to any rights you have to control your personal information and how it's used. Think about those privacy policies you're asked to read and agree to when you download new smartphone apps" [5]. Here is another definition of privacy: "Privacy is often defined as having the ability to protect sensitive information about personally identifiable information, while protection is really a security component. Others define it as the right to be left alone" [6]. Whatever the definition, privacy is an important part of our lives. Is the definition changing as a result of our data-centric world and the Internet of Things?

Some people say that the issue of privacy is not a problem for us as individuals if you are doing nothing wrong or immoral that you are trying to hide. Maybe not, until your right to private data is stolen as a result of a hack or breach. Privacy should matter to every one of us, as it is a constitutional right under our state and federal constitutions. A quick Google search will show you the tremendous concerns of every business in the world.

Privacy is a concern worldwide. Just as an example of many new laws are being enacted as a result of concern for individual privacy, in August 2017, the Supreme Court of India ruled that Indian citizens have a fundamental right to privacy [7]. This is being considered one of India's most progressive judgments that will ensure that the citizens of India will be able to question the government's action. They can now sue the government if they think their privacy is being violated. The effect of this new law will no doubt result in more new laws and regulations for its own agencies to protect consumer interests, new corporate policies, etc.

According to the law firm Hunton Andrews Kurth LLP, an international law firm [8], a draft data protection bill (The Bill) has also been proposed in India, which would establish requirements for the collection and processing of personal data, including particular limitations on the processing of sensitive personal data and the length of time in which personal data may be retained. The Bill would require organizations to appoint a Data Protection Officer and require annual third-party audits of the organization's processing of personal data. The Bill would also require organizations to: (1) implement certain information security safeguards, including (where appropriate) de-identification and encryption, as well as safeguards to prevent misuse, unauthorized access to, modification, disclosure or destruction of personal data; and (2) require regulator notification and, in certain circumstances, individual notification in the event of a data breach. Noncompliance with the Bill would result in penalties up to 50 million Rupees (approximately USD $728,000), or 2% of global annual turnover of the preceding financial year, whichever is higher. Similar to these regulations is the GDPR in the European Union, which carries hefty fines for violations of the regulation.

What is Security? Well, to start, you cannot have privacy without security. Security relates to how your personal information is managed and protected, by you and others who manage your personal data. As we increasingly use the Internet, we learn how our personal information is stored in many different locations, including the applications you download on your computer or smart phone. You are required to download your information to use the app. What you may not realize is how that data is stored, used and sold. Security protects us from unauthorized users. Our IT professionals put those security controls in place to determine who can access data and how. Because of the IoT, cybersecurity is no longer just an issue for IT professionals who maintain and control the security of information. It is now, and should be, one of the most significant issues to all C-suite officers and their board of directors who are ultimately liable for the top decisions made for the company. Some board members have been the subject of significant litigation resulting from a breach of fiduciary duty on these matters, as discussed later in this chapter.

2.3 Managing Privacy and Security: Whose Responsibility Is It?

Obviously, the responsibility lies with all of us—as individuals and as companies. When you download an app, there are certain precautions you must take. Malicious apps ask you to download unnecessary information that may signal that it is acting as a backdoor to your device. According to a study of more than 400,000 apps available from the Google Play store by cyber security company, NowSecure, 10.8% of all apps leak sensitive data over the network, 24.7% of mobile applications have at least one high-risk security flaw, and 50% of popular apps send data to an ad network including but not limited to phone numbers, IMEI numbers (a unique identifier assigned to cellular devices), call logs, and location coordinates [9].

As companies, there is also the ethical and legal responsibility of safeguarding the personal data of others. Compliance with the myriad laws and regulations takes the work of specialists in the legal and cybersecurity world. We also rely on our politicians to enact laws to protect our data. When laws are enacted we then rely on the judicial system to enforce the law. Due to the high number of breaches, hacking and cyberattacks, individuals and companies can also protect themselves by way of either cyber insurance or technology purchased to prevent hacks and breaches.

What does all this mean for consumers, businesses, and legislatures? While it surely looks like the political landscape is undergoing an evolution of sorts, political action committees and deep pocket businesses will likely continue to discourage new US privacy legislation, but security legislation will gain traction over the next several years. The EU along with other countries will continue to evolve their consumer protection, including comprehensive privacy and security legislation. As can be recognized from a quick look at the new proposed legislation in the USA, the USA is slow to enact more updated privacy laws. As an example, in September 2018, the Senate Committee on Commerce, Science and Transportation held a hearing on data privacy, focusing in part on the potential for federal privacy regulation, which centered on two issues: (1) the potential for Congress to pass a privacy law, including the scope and model for any such law and (2) the role of the Federal Trade Commission (FTC) in regulating data privacy practices [10].

As stated by Secureworks, US consumers will need to be more diligent in protecting themselves. We all know few read the terms and conditions when it comes to apps and websites to protect their own privacy, consumers might need to be more discerning regarding to whom they grant access to their information. Consider security implications and data encryption options when making purchases that store personal data. Businesses can elect to adopt privacy by design (e.g., capture less data) and security by design methodologies (e.g., use strong passwords that must be changed by the consumer), not because it is lucrative or required by statute, but because it proactively helps protect consumers. Perhaps this will be seen as a compelling market differentiator, but until such time, expect the proliferation of IoT and big data to continue unabated. Data stores will grow exponentially as will the sophisticated data correlation algorithms [11].

Clearly, with all the hacking and breaches that occur, whether actual security exists is a fair question. Consider this statement by Randy Manner and Brian D. Walker on the NACD's Board Talk [12], "Like the private sector, there are plenty of federal government examples of failing in the fight: the loss of sensitive data on 22 million individuals by the Office of Personnel Management, the hacking of the Chief of Staff to the President, and the loss of highly sensitive cyber defense tools by the NSA are but a handful of examples. Bottom Line? If the government struggles to defend itself, it can't be expected to defend businesses." The message here is managing privacy and security is the responsibility of every individual, every company, and every government agency. None of us can do it alone. Hackers work 24 h a day, making attempts to steal our data and sell it to the unscrupulous buyers to our serious detriment.

2.4   Balancing and Managing Conflict Between Privacy and Security

When considering this issue, ask yourself whether you are willing to give up some of your privacy to improve your safety. For example, do you mind cameras watching you at every street corner, with the ability to zoom in and apply facial recognition techniques presumably only to catch criminals? What happens to that data may be out of your control and may be used for other larger purposes. These are issues we must address with no easy answer. In the USA, we are acutely aware of the frequency of terrorist attacks and killings, the latest one as of this writing, on November 7, 2018 in southern California. It is happening everywhere in the world. On the one hand, we criticize Facebook and other forms of social media for violating our privacy as well as allowing radicalization and propaganda activities. On the other hand, Americans love social media and are generally careless with our use, making us an easy target for hackers. As of the third quarter of 2018, Facebook had 2.27 billion monthly active users worldwide.

2.5   Cost of Cyberattacks on Small- and Medium-Sized Firms

In 2017, cyberattacks cost small and medium-sized firms an average of $2.2 million and are reaching a new level of sophistication [13]. The Ponemon Institute reports that the average cost for small businesses to clean up after being hacked is about $690,000 and, for middle market companies, it is over $1 million [14]. The US National Cyber Security Alliance reports that 60% of small companies are unable to sustain their business more than 6 months following a cyberattack. According to Verizon's 2018 Data Breach Investigations Report [15], 58% of all cyberattacks target small businesses. The reason for that is small businesses do not have the financial resources to buy the most technologically advanced protection. This statistic makes it clear that the cybercriminal is not always looking for that huge financial payout or huge amount of personal data to sell. All it takes is a single unprotected or improperly secured device or careless employee to access an entire system. A good example is the LabMD case, which resulted from a single employee's actions.

These numbers are mentioned because as engineers, the IoT products must be developed keeping in mind their ability to be updated with new software to prevent breaches. A lack of attention to this concern could financially drain a company. LabMD went out of business as a result of litigation with the Federal Trade Commission [16] and in a lawsuit that has lasted several years, had a narrow success on appeal. LabMD took the strong position that the system it had in place to secure personal medical information of patients was sufficient. The FTC did not see it that way. LabMD is now petitioning the court [17] for $1.8 million in attorney's fees and costs [18]. The litigation costs for this issue alone will be high. The point is that compliance is a choice, but the risk is great.

In a company, no matter the size, the commitment required to secure personal information is critically necessary and must involve everyone from the top to the bottom. Senior executives must have a strong understanding of the data at risk, where it is safeguarded, who has access to it and how every single employee is trained to secure the information. Small companies should pay even closer attention to the cybersecurity risks because they seem to be more lax at protecting their data [19]. One reason may be because they believe they have less to lose than the big corporations. But actually, they may not be able to survive the fallout or cleanup due to lack of security or lack of compliance.

2.6 Compliance v. Security Are they One and the Same?

Think about it this way—Compliance is a demonstration, a report, usually mandatory, of how your security program meets specific security standards as laid out by regulatory organizations such as the Securities and Exchange Commission (SEC), NIST, Cybersecurity Act of 2015, HIPAA, the European Union's GDPR, or even the Sarbanes-Oxley Act of 2002, to name a few [20]. With the voluminous enactment of cybersecurity laws, it is an arduous job to keep up with them and most of time, the laws are constantly attempting to keep up with technology. In an article by Techspective.net, here is their perspective on what the USA is doing to keep technology safe: "Technology progresses rapidly. The government, infamously, does not. It takes a long time to pass new laws, making it difficult for cybersecurity to keep up to date with changing technologies. As of now, many new forms of technology are not tightly regulated" [21]. Consequently, new laws are implemented and current laws are updated frequently, making compliance difficult. Many regulatory agencies send out press releases giving directives to public companies. For example, in a press release by the SEC on October 16, 2018, public companies were cautioned to consider cyber threats when implementing internal accounting controls. This press release was a result of the SEC Enforcement Division's investigations of several public companies who fell victim to cyber fraud, and consequently losing millions of dollars in the process.

2.7 Local, State, Federal, and International Laws

As can be seen in the writings of this chapter, cyber legislation is alive and well locally, nationally and internationally. Below are a few of the powerful acts of legislation currently in place. This list is far from exhaustive, as new legislation on the topic of cybersecurity seems to be on the top of every legislator's list. Keep in mind that certain law firms specializing in cybersecurity law keep active lists of all cyber-security legislation.

Legislation is enacted to protect the security of our information and it usually has a compliance side to ensure that penalties are in place for noncompliance with the legislation. Many countries have now enacted similar laws. Local state bar associations such as the Colorado Bar Association continually provide legal guidance for lawyers in small, medium-sized and large law firms attempting to calm down the panic caused by its enactment and to provide compliance models.

2.7.1 The Clarifying Lawful Overseas Use of Data Act (CLOUD)

The Clarifying Lawful Overseas Use of Data Act or CLOUD Act (H.R. 4943, 115th Congress 2017–2018) is a US federal law enacted in 2018 by the passing of the Consolidated Appropriations Act, 2018, PL 115-141, section 105 executive agreements on access to data by foreign governments. Primarily the CLOUD Act amends the Stored Communications Act (SCA) of 1986 to allow federal law enforcement to compel US-based technology companies via warrant or subpoena to provide requested data stored on servers regardless of whether the data are stored in the USA or on foreign soil [22]. This Act resulted from difficulties the Federal Bureau of Investigation (FBI) had during a criminal drug traffic investigation. Some of the emails of a US citizen were stored on a Microsoft remote server in Ireland, which Microsoft refused to honor pursuant to a warrant issued by the FBI. This led to the FBI filing a lawsuit against Microsoft Corporation, which was finally resolved by the US Supreme Court in *United States v. Microsoft Corp.* [23].

2.7.2 The Cybersecurity Information Sharing Act of 2015 (CISA)

On December 18, 2015, President Obama signed into law the Cybersecurity Act of 2015. This significant piece of federal cyber-related legislation establishes a mechanism for cybersecurity information sharing among private-sector and federal government entities. The Cybersecurity Act of 2015 is Division N of the omnibus spending bill. Congress passed this Act to increase the sharing of cybersecurity information among businesses and between businesses and the government and to improve the quality and quantity of timely, actionable cybersecurity intelligence in the hands of the private sector and government information security professionals.

Among many other specific directives, this Act requires the heads of certain governmental agencies, to jointly develop and issue procedures to facilitate and promote timely sharing of classified cyber threat indicators and defensive measures in the passion of the Federal Government with representatives of relevant federal entities and nonfederal entities that have appropriate security clearances. It also establishes privacy and civil liberties guidelines governing the receipt, retention, use and dissemination of cyber threat indicators (CTI) and defensive measures (DM) by a federal entity under CISA.

2.7.3 National Institute of Standards and Technology (NIST)

The National Institute of Standards and Technology (NIST) was founded in 1901 and is now part of the US Department of Commerce. NIST is one of the nation's oldest physical science laboratories. Congress established the agency to remove a major challenge to US industrial competitiveness at the time—a second-rate measurement infrastructure that lagged behind the capabilities of the UK, Germany, and other economic rivals. NIST's mission is to promote US innovation and industrial competitiveness by advancing measurement science, standards, and technology in ways that enhance economic security and improve our quality of life [24].

2.7.4 Information Sharing and Analysis Centers, Information Sharing and Analysis Organizations (ISAC, ISAO) [25]

Information Sharing and Analysis Centers (ISACs) were created in response to a 1998 US Presidential Decision Directive signed by President Clinton for federal agencies to encourage industry sectors to establish organizations to share security threat and vulnerability information with critical infrastructure owners and operators and the federal government [26]. ISAOs were established as a result of an Executive Order signed by President Obama in 2015 to enable private companies, nonprofit organizations, federal departments and agencies to share information related to cybersecurity risks and incidents and collaborate to respond in as close to real time as possible [27]. Europe now has ISACs as well [28].

2.7.5 The Internet of Things Cybersecurity Improvement Act

The Internet of Things Cybersecurity Improvement Act was introduced in 2017 by Senators Mark Warner (Democrat, Virginia), Cory Gardner (Republican, Colorado), Ron Wyden (Democrat, Oregon), and Steve Daines (Republican, Montana), and provides more clarity on IoT security standards in the USA. This bill would force vendors who sell technology to the US government to ensure that those devices can receive security patches and not rely on passwords that cannot be changed. It also prohibits the shipment of devices with known vulnerabilities.

2.7.6 European Union General Data Protection Regulation (GDPR)

This law is one of the most important changes in data privacy regulation in the past two decades. Its impact is widespread and worldwide with its extended jurisdiction reaching and applying to all companies anywhere in the world who process personal data of persons residing in the European Union. The purpose of the GDPR is to protect all EU citizens from privacy and data breaches [29]. On the first day of this new law, May 25, 2018, Google and Facebook faced privacy complaints [30].

Organizations in breach of GDPR can be fined up to 4% of annual global turnover or €20 Million (whichever is greater). This is the maximum fine that can be imposed for the most serious infringements, for example, not having sufficient customer consent to process data or violating the core of Privacy by Design concepts. There is a tiered approach to fines. For example, a company can be fined 2% for not having their records in order (article 28), not notifying the supervising authority and data subject about a breach, or not conducting [an] impact assessment. It is important to note that these rules apply to both controllers and processors—meaning "clouds" are not exempt from GDPR enforcement.

2.7.7 The EU Cybersecurity Act

Proposed in 2017 to deal with cyberattacks and promote enhanced cybersecurity in the European Union, it was announced on December 10, 2018 that a political agreement was reached between the European Parliament, the Council of the European Union and the European Commission on the EU Cybersecurity Act [31]. This act would empower the EU's European Network and Information Security Agency (ENISA) to formulate voluntary cybersecurity standards for systems used in power plants, medical devices, connected consumer devices, and other services. Tommy Ross, senior policy director at the Building Societies Association's, Washington D.C office says "The EU approach will be the first large-scale experiment in how to lift the tide in the consumer marketplace when it comes to security." If this act goes into force, ENISA would likely implement standards for secure software development, identity management, and hardware security rules. The European Commission is set to draft the scope of products that require obligatory certification, with a list to be finalized by 2023 [32].

2.7.8 Brazil's "GDPR"

On May 29, 2018, the Brazilian House of Representatives approved a data protection bill (PL52/3028). The next step is with the Senate for analysis and possible approval. This bill requires all companies that treat or aim at Brazilian data generated within the country to be subject to its provisions.

2.8   The Cloud and the Supply Chain: Is It a Good Idea?

Cloud based computing is a method of computing which holds strong potential for streamlined sharing, input and output of data sets among teams, individuals or cross-organization. The Cloud allows users to access applications, information and data online rather than through hardware or devices. Typically, cloud-computing solutions have the capability to be "plugged in," meaning sourcing the data from one cloud to another. There are public and private clouds. The benefits vary according to the type of cloud service, but generally, using cloud services allows companies to not have to buy or maintain their own computing structure. That means a company can forego buying servers, updating applications and operating systems or decommissioning and disposing of hardware or software when it is out of date. That is all taken care of by the provider. One would assume that companies who specialize in this service are likely to have better skills and more experienced staff than that a small business could afford to hire. The point is to use cloud service providers who can deliver a more secure and efficient service to its users.

However, there is a caveat. Consider whether there is accountability, responsibility, and liability in the cloud. A multitude of questions should be asked before signing a contract, including who owns the data. Moving to a particular cloud service may mean that you are using the same applications as a rival, causing difficulties in getting a competitive advantage if that application is core to the business. Think about whether you should be reaching out to the cloud provider for ensuring cloud compliance. There may be differing interests and objectives. Customers want transparency from providers regarding risk factors involved with the cloud. Providers may not want to disclose everything behind their cloud operations. Having a thorough audit strategy will assist with these concerns. There are also many guides to secure cloud computing. Whether you use cloud computing or you are the developer of a cloud service, privacy should always be considered in terms of legal compliance and user trust; privacy needs to be considered at every phase of design.

Unsurprisingly, the USA does not have one regulation or law for data regulation across the country. Instead, there is a dizzying number of cloud computing federal regulations and requirements and state legislation that work together to keep our data safe. Obviously, the difficult part is compliance, thus the need to have a team working together at the production stage.

2.9   Constitutional Implications of IoT

The First Amendment to the US Constitution states: "Congress shall make no law respecting an establishment of religion or prohibiting the free exercise thereof; or abridging the freedom of speech, or of the press; or the right of the people peaceably to assemble, and to petition the Government for a redress of grievances."

Can a portable IoT system help safeguard freedom of speech? Think about this: A person driving an automobile who exercises due care but injures another person who is participating in a protest or demonstration and is blocking the traffic in a public right-of-way is immune from civil liability for such injury but shall not be immune from the same if the actions of the driver leading to the injury were willful or wanton. This is the language of a Rhode Island bill introduced by Representative Justin Price in March 2017 [33]. Similar bills have been introduced in Texas, Florida, Tennessee, and North Carolina. Could bills like this have a chilling effect on free speech? These bills have faced criticism but they could be enacted into law. Is it a good law? Does it affect your first amendment right to free speech? Does it mean that, as a peaceful protestor, one has to protect themselves from oncoming cars during peaceful protests? Can this simple IoT system protect you from harm caused as a result of a law like this?

Whether it seems bizarre to think we need such technology, there is IoT-based protection for protestors being developed. Companies such as FabLab and Honeywell offer this type of technology used to detect cars and their speed. Other technology with light-based sensors uses cameras to capture texture, color, and con-trast information to detect oncoming vehicles. Light Detection and Ranging (LiDAR) sensors such as those from NXP measure the distance to an object by calculating the time taken by a pulse of light to travel to an object and back to the sensor [34]. This technology seems like a drastic step needed in order to protect our first amendment rights but, as we all know, technology moves at lightning speed compared to the legislation process. According to statista.com [35], the number of Android app releases per day in the first quarter of 2018 was 6140. For that same quarter, 1434 mobile apps were released through the Apple App Store every day.

New apps are created every day and whether an idea is patentable is the big ques-tion. In a 2014 decision [36], the US Supreme Court set out a two-step procedure that courts are to follow in determining whether a computer-implemented invention is no more than an abstract idea that is ineligible for patent protection. The patents at issue in this case disclose a computer-implemented scheme for mitigating "settle-ment risk," that is, the risk that only one party to a financial transaction will pay what it owes, by using a third-party intermediary. The question presented is whether these claims are patent eligible under 35 U.S.C. § 101 or are instead drawn to a patent-ineligible abstract idea. The Supreme Court affirmed the judgment of the US Court of Appeals for the Federal Circuit and held that the claims at issue are drawn to the abstract idea of intermediated settlement, and that merely requiring generic computer implementation fails to transform that abstract idea into a patent-eligible invention.

Apparently, the Court's newest instructions on how to determine whether an invention is patent-eligible raise as many questions as they answer. As engineers, this is exactly why you should consider having lawyers involved at the early stages of product development.

2.9.1 The IoT and the Fourth Amendment to the US Constitution

The Fourth Amendment states "The right of the people to be secure in their persons, houses, papers, and *effects* (emphasis added), against unreasonable searches and seizures, shall not be violated, and no Warrants shall issue, but upon probable cause, supported by oath or affirmation, and particularly describing the place to be searched, and the persons or things to be seized." Why I italicized the word "effects" will be explained below.

Does the fourth amendment apply to the Internet of Things? Did our forefathers anticipate or consider the IoT? This is a hot legal issue. What constitutes "effects" as stated in the fourth amendment is a newer legal concept that has come to light partly as a result of the IoT. The term "effects" has long been understood to signify the protection of personal property, including a vehicle [37]. So it seems appropriate to apply the fourth amendment to gadgets, devices, etc. that communicate through the Internet of Things. It sounds simple, but it gets very complicated.

In a US Supreme Court case, *United States v. Jones* [38], the Government obtained a search warrant permitting it to install a global positioning system (GPS) tracking device on a vehicle registered to Jones's wife. The warrant authorized installation in the District of Columbia within 10 days, but agents installed the device on the 11th day and in Maryland. The Government then tracked the vehicle's movements for 28 days. By means of signals from multiple satellites, the device established the vehicle's location within 50–100 ft and communicated that location by a cellular phone to a Government computer. It relayed more than 2000 pages of data over the 4-week period. The Government subsequently secured a criminal indictment of Jones and others on drug trafficking conspiracy charges. The District Court suppressed the GPS data obtained while the vehicle was parked at Jones's residence but held the remaining data admissible at trial because Jones had no reasonable expectation of privacy when the vehicle was on public streets. Jones was convicted. The D.C. Circuit reversed the conviction, concluding that admission of the evidence obtained by warrantless use of the GPS device violated the Fourth Amendment. The Supreme Court held that the Government's physical intrusion on an "effect" (in this case, a vehicle) for the purpose of obtaining information constitutes a "search."

2.9.2 Fifth Amendment Implications

The Fifth Amendment to the US Constitution States the following: No person shall be held to answer for a capital, or otherwise infamous crime, unless on a presentment or indictment of a grand jury, except in cases arising in the land or naval forces, or in the militia, when in actual service in time of war or public danger; nor shall any person be subject for the same offense to be twice put in jeopardy of life or limb; nor shall be compelled in any criminal case to be a witness against himself, nor be deprived of life, liberty, or property, without due process of law; nor shall private property be taken for public use, without just compensation.

The focus here will be the right to due process as it relates to your personal data and the IoT. The IoT is alive and well in the courts. Just think of the massive amounts of devices, gadgets and accessories we all use every day. Each of these devices includes sensors and microprocessors that connect with the IoT. This is no longer just mobile phones. As mentioned above, we all use IoT devices, be it your iWatch, your home security system, a Fitbit, a thermostat, a toy, or maybe medical equipment that connects to the IoT.

IoT use is prevalent and clearly affects how our world is changing before us. The massive amounts of data available through the IoT are mind-boggling. Information gained from IoT devices is being used in the courts every day. Consider the personal injury case where a law firm is using a woman's Fitbit to prove she suffered from reduced ability to be active following an accident. In this case, the information may benefit the injured party. The bigger question is who can legally gain access to this type of information and how can it ultimately be used against you in a future lawsuit. According to Forbes, "the lawyers aren't using Fitbit's data directly, but pumping it through analytics platform Vivametrica, which uses public research to compare a person's activity data with that of the general population" [39]. Does access to that information require your specific consent?

2.9.3 Equal Protection Clause of the 14th Amendment to the US Constitution

Section 1 of the Fourteenth Amendment states as follows: All persons born or naturalized in the USA, and subject to the jurisdiction thereof, are citizens of the USA and of the state wherein they reside. No state shall make or enforce any law which shall abridge the privileges or immunities of citizens of the USA; nor shall any state deprive any person of life, liberty, or property, without due process of law; nor deny to any person within its jurisdiction the equal protection of the laws.

Let us consider the article written by Lauren Smith entitled "Unfairness by Algorithm: Distilling the Harms of Automated Decision-Making" [40]. Whether obtained illegally or legally, information collected by way of the Internet can be used in an unfair way—a way that discriminates against certain individuals. Of course, we know that there are ways in which data collected can improve services, advance research, and combat discrimination. But this same data and analysis can also have a detrimental impact on vulnerable communities, especially when automated decision-making uses sensitive data such as race or gender. This issue is addressed in Lauren Smith's article. It highlights legal and ethical issues raised by using sensitive data for hiring, policing, benefit determinations, marketing, and other purposes [41]. In their research, they distilled both the harms and potential mitigation strategies identified in their literature and presented to the FTC for consideration for the FTC Informational Injury workshop for use in assessing fairness, transparency, and accountability for artificial intelligence, as well as methodologies to assess impacts on rights and freedoms under the EU General Data Protection Regulation.

They found that potential individual and collective/societal harm could occur, with some individual harm being illegal and some unfair. For example, filtering job candidates by race or genetic/health information could lead to race discrimination or filtering candidates by work proximity could lead to excluding minorities. Another example would be use of recidivism scores to determine prison sentence length (leading to a societal harm of disproportionate incarceration) or differential pricing of goods and services (i.e., raising online prices based on membership in a protected class). The bottom line is that automated decision-making can lead to individual and societal harms in many areas such as employment, housing, insurance and benefits, and education.

Again, the legal issues are being presented to you to consider having a lawyer on your team at the product development stage.

2.10 It Is All About Survival: How to Protect Your Privacy and Security as Individuals

As individuals, when downloading an app to a smart phone, we are asked if we consent to the use of our information. The consent given can be very broad. Take the time to read and understand what consent you are giving and to whom. It appears that only those individuals who actually read the agreement are the professionals who are researching or studying the topic. Most people just do not do it. What happens if you do not give consent? Will you still be able to download the app, or get your information processed or compared for your benefit? Most likely not, but at least you know what you are doing and that you can ultimately do something about it.

We should be involved in effecting change in legislation currently being proposed to protect our privacy and security. Even most smartphone consent agreements have multiple sections and pages. I have asked hundreds of people whether they read the entire agreement before making the decision to give consent. Only one, a presenter at a cybersecurity conference, said he reads the entire agreement. My review of my own Netflix agreement is multiple pages with several sections. It tells you how to opt in or out of certain services. It is important to read the agreements provided and understand what you are allowing to be done with your information.

For some smart gadgets, it is not as easy for most users but it is important to go to the app's website and review privacy preferences in account settings and control how your information can be used. It would be prudent to do business with companies and organizations that value your privacy and take measures to protect your personal information. There are simpler things you can do too, to help protect your privacy and boost your security.

Here are some suggestions:

- Limit what you share on social media and online in general.
- Shred important documents before tossing them in the trash.
- Guard your Social Security number. Keep it in a secure place and do not give it out if possible. Ask if you can provide another form of identification.
- Safeguard your data and devices. This might include enlisting the help of computer virus protection, a secure router, Wi-Fi protection, and identity theft protection services.
- Get a VPN (Virtual Private Network).
- Understand how the information you are giving away could be used. Read an organization's privacy policy before signing up for an app or service. Understand exactly what you consent to when downloading an app.
- Remember, if the app or service is free, the company may make its money by selling your data.

As engineers, avoidance of the possibility of having to correct big mistakes later in production is critical and could make or break the company. Having that diverse team involved at the product development stage is something to seriously consider. That team should include legal, compliance and cybersecurity specialists with whom the engineers can discuss product development from a different but important perspective. Engineers and developers should be asked what they are doing or would do to anticipate and mitigate not only particular cybersecurity threats but also ongoing compliance concerns, especially considering ongoing changes in the law. Businesses cannot afford to be reactive anymore.

2.11   Emerging Cybersecurity Threats in 2018 and 2019

Cybersecurity threats are constant and powerful. Hacking via the Internet is prevalent. Big cyberattacks, like the attack of the Equifax credit reporting agency in 2017, led to the theft of significant personal data on almost half of the US population [42]. Even more recently, hacking has occurred of public Wi-Fi in a Starbucks in Argentina [43]. Chinese hackers allegedly stole data of more than 100,000 US Navy personnel, resulting in an indictment issued on December 20, 2018 by the US Justice Department. By sneaking into the computers of a US company that manages IT systems remotely for other businesses, the Chinese hackers were allegedly able to access computers at more than 45 companies in a dozen countries including the USA, the UK, Australia, India, and Japan [44]. No one individual or company, big or small is safe from these types of attacks. Companies that hold the most sensitive of data are prime targets, especially health information. Marc Goodman, a security expert and author of Future Crimes, thinks data brokers who hold information about things such as people's personal Web browsing habits will be especially popular targets. He says, "These companies are unregulated, and when one leaks, all hell will break loose" [45]. According to Martin Giles, one big target will be the cloud computing businesses, as they house huge amounts of data for companies, including emails and photo libraries. He believes that smaller companies are likely to be the most vulnerable, because they do not have funds to hire the brightest minds in digital security, as do Google, Amazon, and IBM. But we all know that big companies have had huge breaches as well compromising our personal information.

Other anticipated future targets are electrical grids, transportation systems, older planes, trains, ships, and other modes of transport that could [46] leave themselves vulnerable. Also see, "Hijacking Computers to Mine Cryptocurrency is all the Rage" [47]. Apparently, the Showtime Website contained a tool that was secretly hijacking visitors' computers to mine Monero, a Bitcoin-like digital currency focused on anonymity. Other predictions include nations will be at cyberwar, supply-chain attacks will rise, cybersecurity will raise its profile in the corporate board room, and small and mid-size business will find ways to monitor and detect threats and respond when necessary, finally realizing that they are just as big a target.

2.11.1   Evolving Tools and Strategy

With every day that passes, the news reflects more concern about threats becoming more complex. Looking back at the job of the head of an IT department, it was much simpler, at least in terms of the number of devices they needed to protect, especially when compared to technological advances of today. Now, the ways in which cybercriminals gain access to enterprise networks makes the job much more complicated and security teams must develop new tactics to fend off the advanced threats levelled against the increasingly interconnected enterprise networks. These concerns are today a concern of the entire company, that is, board of directors, not just the IT department, as a result of increasing cyberbreaches, putting cybersecurity at the top of the agenda for board members. Attacks such as WannaCry and NotPetya [48] ransomware outbreaks have finally caught the attention of most boards and small business owners even if they have not yet been attacked. Security is now a prerequisite that is built into new technologies and products from the outset. The sad part is that according to Verizon's 2018 Data Breach Investigations Report, it shows that the same threat tactics are still effective in infiltrating data, because many organizations are missing a core foundation of security tools and processes [49].

2.12   Where Is the Board? The Board's View of Cybersecurity

Board directors are now becoming quite aware of their fiduciary duty to ask questions about cybersecurity risks and the liability they have undertaken. The former officers and directors of Yahoo agreed to pay $29 million to settle charges that they breached their fiduciary duties in their handling of customer data during a series of cyberattacks from 2013 until 2016 [50]. In a derivative lawsuit, which is the legal path for shareholders to hold corporate directors and top management accountable for their actions, this represents the first time that shareholders have been awarded monetary damages related to a data breach [51]. Even before this settlement occurred on April 24, 2018, the Securities and Exchange Commission (SEC) had already severely sanctioned Altaba (which was Yahoo at the time) with a $35 million penalty for failing to make timely disclosure of the data breach [52].

The fiduciary duties of a board director should not be taken lightly. Huge liability rests on the shoulders of each board member. Board directors should be asking what their company's plan is if a breach were to occur. They should be inquiring of product engineers and developers what they are doing in anticipation of and to mitigate particular threats. According to an article by Katie Swafford, "Directors… could sit together and say, 'All right, we know that this happened to a competitor, or we know that this is a trend. How would we as an organization handle it?' And they can ask their management to report back on what the company would do in this case" [53]. If board members don't know the answer, as Ms. Swafford states in her article, "If your board runs through this exercise and struggles to find answers, fear not—a plethora of consultants, experts, companies, and independent resources are eagle to help."

In the case of *Reiter v. Fairbank*, [54], Capital One Financial Corporation's eight board of directors were sued in a derivative action where a shareholder asserted that they breached their fiduciary of loyalty and unjustly enriched themselves by consciously disregarding their responsibility to oversee Capital One's compliance with the Bank Secrecy Act and other anti-money laundering laws. The defendants were successful in getting this case dismissed. The Court found that the plaintiff failed to allege facts from which it could be reasonably inferred that the defendants consciously allowed Capital One to violate statutory requirements so as to demonstrate that they acted in bad faith. In an article written by Francis G. X. Pileggi, he states, "Although the company settled without admitting liability, Capital One admitted that it failed to adopt and implement a compliance program that adequately covered various money laundering programs due to an inadequate system of internal controls and ineffective independent testing." Michael Reiter, a shareholder of Capital One, subsequently sued the company's board members alleging that the settlement was evidence of the board's failure to fulfill its fiduciary duty of oversight. As Pileggi states, "*Reiter v. Fairbank* provides a practical lesson that outlines a director's duties of oversight and defines the conditions under which a plaintiff can successfully prevail" [55]. The lesson to be learned is that if board directors consciously disregard their fiduciary duties, they could be held personally liable. For this reason, it is critical that board members be aware of what is going on at all levels, including the product development level to ensure compliance.

In addition to the concerns regarding internal cyber-risk assessment by the board, the board must not forget its responsibility of cyber-risk oversight duties as it relates to the company's third-party cybersecurity risk management programs. Third-party companies who have access to a company's sensitive date could be vulnerable, as they have different security systems. Board members should be asking many questions. Here are a few: What does the company do with the information after they obtain it? Who has access to it? How is it stored? Do any subcontractors have access to it? What is the firm's information technology security policy? What is their policy and time frame in which the professional services firm will alert the company should a breach occur? Do they think about who uses the data and alter its cybersecurity inquiries accordingly? These are critical questions that should be considered and asked by any board of directors, as their fiduciary duty.

2.13   Cyber Insurance

Until recently, cyber insurance was not popular and either not considered or purchased. It is now very common. As a member of a local quasi-governmental organization, I raised the issue of cyber insurance for my agency. I was told that the greater organization of which we were a part, had considered cyber insurance but had no budget for it. In fact, they applied for cyber insurance but could not qualify because they could not answer questions to the satisfaction of the insurance company. Back then, this scenario seemed to be fairly common, based on my conversations with other agencies. Go online now and you will find an onslaught of ads from insurance companies ready and willing to sell you cyber liability insurance. Adam Sandler, head of cyber solutions at RMS, recently said: "RMS clients are seeing demand for cyber insurance growing rapidly and their ability to pursue this opportunity is constrained by their ability to allocate risk capital with confidence" [56]. The decision to purchase cyber insurance is a matter of weighing the cost versus the risk. Some companies believe they can self-insure and cover that risk themselves. It is a decision that needs to be carefully considered.

2.14   Conclusion

Knowing that cybersecurity laws are enacted every day, it goes without saying that you, as engineers and developers, will be required to collaborate, align, and work with in-house and outside counsel during the product development phase. Doug Hall, founder and CEO of the Eureka! Ranch and the Innovation Engineering Institute, calls for everyone, from leaders to the front line, to think smarter, faster, and more creatively. Doug Hall's concept of innovation engineering reimagines how innovation is led, managed, and delivered. He calls it a new field of academic study and leadership science. In his book, Innovation Engineering® The Art and Science of System Driven Innovation™, Mr. Hall teaches how to apply the science of system thinking to help us work smarter, faster, and more creatively. It transforms innovation from a random gamble to a reliable science. Mr. Hall boasts that Innovation Engineering is reliable because it is grounded in data, backed by academic theory, and validated in real world practice. He states that collectively Innovation Engineering is the number-one documented innovation system on earth. Over 35,000+ people have been educated in Innovation Engineering classes and, over $75 billion in documented growth and system improvement projects are currently in active development [57].

Should product engineers be aware of potential problems with their product as it relates to privacy and security? Some would argue no, that it stumps creativity. Others argue that the engineers have an ethical responsibility to consider privacy and security concerns. As can be recognized from the types of lawsuits filed relating to privacy and security, it could prevent lawsuits if privacy and security is considered at the product development stage.

Have you heard of a "second Internet"? You can find out about it online, but here is a preview from Professor Howard. Some experts believe the "IoT need not be a threat to personal privacy and political freedom and that it can be regulated through a second Internet. The next Internet is still far enough away that citizens can have a voice in how it is constructed and operated: to create systems, rules, and regulations that ensure that the Internet of Things will serve people, not betray them. But such a Constitution must be created from the ground up" [58]. Clearly, we all like the convenience information gathering provides and the benefits that can result from this aggregate information. Can it be misused? Yes. As Professor Howard states, that is why we as citizens and leaders must realize what is at stake and at risk if action is delayed. "Awareness of cybersecurity and the risks of having poor or no security is on the rise but many companies still haven't taken action, as a survey shows 44% of the 9,500 executives who were asked said their organization doesn't have an overall information security strategy" [59]. A good example of the impact of international laws is Apple opening a data center in China to comply with their new cybersecurity law. This decision is a result of China's new law that requires companies to store users' data in China. US corporate compliance with international cybersecurity laws is required to avoid extremely costly fines for noncompliance. Microsoft, Amazon, and Facebook are among the big American technology companies plowing billions of dollars into building data centers in Germany, the Netherlands, France, and other countries [60].

A last thought—Keep in mind that according to Verizon's Data Breach Investigations Report, 76% of breaches were financially motivated, which means that if there is some way possible money can be taken from you, by whatever means, that is, credit card data, personally identifiable information, intellectual property, driver's license information, credit card applications, whatever you have that is valuable, it will happen to you. So be prepared, be vigilant, and keep security in mind at the product innovation stage, hopefully with a team concept approach, which has been advocated throughout this chapter.

References

1. Kranzberg M (1986) Technology and history: "Kranzberg's Laws". Technol Cult 27(3):544–560. https://doi.org/10.2307/3105385. Retrieved from http://www.jstor.org/stable/3105385
2. https://www.economist.com/leaders/2017/05/06/the-worlds-most-valuable-resource-is-no-longer-oil-but-data
3. https://www.accenture.com/t20160602T030231__w__/us-en/_acnmedia/Accenture/Conversion-Assets/DotCom/Documents/Global/PDF/Dualpub_19/Accenture-Provider-Cyber-Security-The-$300-Billion-Attack.pdf
4. https://www.fbi.gov/wanted/cyber/iranian-mabna-hackers
5. https://us.norton.com/internetsecurity-privacy-privacy-vs-security-whats-the-difference.html
6. https://www.secureworks.com/blog/privacy-vs-security
7. https://fossbytes.com/right-privacy-fundamental-right-cyber-security-impact-india/.   2017 Aug 29
8. https://www.huntonprivacyblog.com/2018/07/27/indias-draft-data-privacy-law-issued-today/
9. Linkedin.com. 2018 June 7
10. Inside Privacy. 2018 Oct 1
11. Privacy vs. Security. Do today's models work with the Internet of Things and its cousin, big data? Thursday, 2017 March 23. By: Brian Dean. https://www.secureworks.com/blog/privacy-vs-security
12. https://blog.nacdonline.org/posts/the-cyber-blind-spot
13. WSJ Pro Cybersecurity. 2018 Sept 18
14. The Denver Post Published, 2016 October 23 at 12:01 am, Updated 2017 March 24 at 12:29 pm
15. Enterprise.verizon
16. The book, The Devil Inside the Beltway: The Shocking Expose of the US Government's Surveillance and Overreach Into Cybersecurity, Medicine and Small Business, by Michael J. Daugherty
17. https://www.ftc.gov/system/files/documents/cases/labmd_ca11_ftc_opposition_to_fee_request_2018-1119.pdf
18. LabMD Inc. v. Federal Trade Commission, on Petition for Review of an Order of the Federal Trade Commission (FTC Docket No. 9357)
19. Business.com/Security/Last modified. 2018 Mar 29
20. H.R. 5069, 114th Congress (2015–2016) the Cybersecurity Systems and Risks Reporting Act will amend SOX to also apply to cybersecurity systems and cybersecurity systems officers and bring it up to date
21. https://techspective.net/2018/09/04/what-is-the-us-doing-to-keep-technology-safe/
22. En.wikipedia.org
23. United States v. Microsoft Corp., 584 U.S. ___, 138 S.Ct. 1186, 200 L.Ed.2d 610 (April 17, 2018).
24. Nist.gov
25. https://www.isao.org/information-sharing-groups/
26. https://www.nationalisacs.org/about-isac
27. https://obamawhitehouse.archives.gov/the-press-office/2015/02/13/executive-order-promoting-private-sector-cybersecurity-information-shari
28. https://www.enisa.europa.eu/publications/information-sharing-and-analysis-center-isacs-cooperative-models

29. https://eugdpr.org/the-regulation/gdpr-faqs/

30. CNBC.com. 2018 July 11

31. Governmenteuropa.eu. 11 Dec 2018

32. Euractiv.com. 2018 Dec 10 (updated 2018 Dec 18)

33. H5690 March 2017, Rep. Justin Price, Rhode Island

34. Lidar is commonly used to make high-resolution maps, with applications in in geodesy, geomatics, archaeology, geography, geology, geomorphology, seismology, forestry, atmospheric physics, laser guidance, airborne laser swath mapping (ALSM), and laser altimetry. The technology is also used in control and navigation for some autonomous cars. Wikipedia

35. https://www.statista.com/statistics/276703/android-app-releases-worldwide/

36. Alice Corporation Pty. Ltd., Petitioner v. CLS Bank International et al., 134 S.Ct. 2347 (2014 June)

37. See United States v. Jones, 132 S.Ct. 945, 565 U.S. 400, 181 L.Ed.2d 911, 80 U.S.L.W. 4125, 23 Fla.L.Weekly Fed. S 102 (2012. Criticized 732 F.3d 187 (3rd Cir. 2013), 12-2548, United States v. Katzin

38. 565 U.S. 400 (2012), 132 S.Ct. 945, 181 L.Ed.2d 911

39. Forbes.com. 2014 Nov 16

40. December 11, 2017, Future of Privacy Forum, fpf.org

41. Unfairness by algorithm: Distilling the Harms of Automated Decision-Making, Lauren Smith, 2017 December 11, Future of Privacy Forum

42. https://www.cnet.com/news/equifaxs-hack-one-year-later-a-look-back-at-how-it-happened-and-whats-changed/

43. https://www.cnbc.com/2017/12/12/starbucks-customer-laptops-hacked-to-mine-cryptocurrency.html

44. MIT Technology Review, Technologyreview.com, The Download, 2018 Dec 20

45. MIT Technology Review, Technologyreview.com, Six Cyber Threats to Really About in 2018, by Martin Giles. 2018 Jan 2

46. Forbes Technology Council, Brian NeSmith, 2018 December 28, Cybersecurity Predictions for 2019. https://www.forbes.com/sites/forbestechcouncil/2018/12/28/cybersecurity-predictions-for-2019/#22a3c8bb4a27

47. MIT Technology Review, Hijacking Computers to Min Cryptocurrency is all the Rage. https://www.technologyreview.com/s/609031/hijacking-computers-to-mine-cryptocurrency-is-all-the-rage/, Mike Orcutt. 2017 Oct 5

48. Wired.com, Andy Greenberg Security. 2018 Aug 22

49. Verizon 2018 Data Breach Investigations Report

50. https://altaba.gcs-web.com/static-files/346e8981-1015-49dd-b616-66756ba99173

51. https://www.nytimes.com/2019/01/23/business/dealbook/yahoo-cyber-security-settlement.html

52. https://www.sec.gov/litigation/admin/2018/33-10485.pdf

53. Enhancing Cyber-Risk Oversight by Katie Swafford, NACD Directorship May/June 2017

54. 2016 WL 6081823. decided 2016 Oct 18

55. https://www.eckertseamans.com/app/uploads/PileggiDelawareWatchJanFeb2017.pdf

56. Rms.com, RMS Releases Industry's First Probabilistic Cyber Risk Model, Newark, Calif. 2018 Mar 7

57. Innovation engineering gets its inspiration from Dr. W. Edwards Deming. For additional information on Dr. Deming's work go to www.deming.org

58. Is the Internet of Things Your New Constitution? Philip N. Howard, pnhoward@uw.edu, Professor, Oxford University and University of Washington 9/11/2015. https://www.sas.upenn.edu/andrea-mitchell-center/sites/www.sas.upenn.edu.dcc/files/Howard%20DCC%20Paper%20v3.pdf

59. Wall Street Journal 2017 October 20

60. https://www.nytimes.com/2017/07/12/business/apple-china-data-center-cybersecurity.html

Chapter 3
Privacy in the New Age of IoT

Qi Pan

3.1 Introduction

Our daily lives are becoming increasingly digitised. The line between our offline and online presence is blurring as consumers and organisations rely on digital devices to stay connected and efficient. As a result, the concept of privacy, defined by the Oxford English Dictionary as "a state in which one is not observed or disturbed by other people", is shifting. The preferences of individuals are tracked, health data is monitored, and all of this data can be collected and mined by organisations who monetise data through targeted advertising and third party sharing, and governments who track individuals. The Internet of Things (IoT) is defined as systems of sensors and actuators connected by networks to computing systems, and it relies on a backbone of connectivity and interoperability. As Hugh Durrant-Whyte stated at the Royal Society conference entitled "IoT: Opportunities and threats" on 3 October 2017, the main impact of IoT is in data and how it is used, not in the physical devices which can become obsolete over time [1]. In the past, data was captured and transferred freely from consumers, often in order to receive targeted offers or an improved credit rating. The control of the data lay with the organisations who freely shared this data. However, in the aftermath of the Facebook/Cambridge Analytica scandal, news of which broke in March 2018, light has been shed on the misuse of personal data and the breach of privacy by organisations consumers entrust with their data [2].

In the context of IoT, there are two major risks associated with privacy. The first and more pertinent risk is the privacy associated with consumer and employee data captured from IoT sensors in devices such as wearables, virtual assistants, such as Alexa, and smart cars. The second risk is related to the proprietary information held

Q. Pan (✉)
London, UK
e-mail: qi.pan@hotmail.co.uk

© Springer Nature Switzerland AG 2019
F. D. Hudson (ed.), *Women Securing the Future with TIPPSS for IoT*, Women in Engineering and Science, https://doi.org/10.1007/978-3-030-15705-0_3

by organisations, which can be compromised in the event of a data breach. The impacts of the lack of privacy on individuals and companies are wide-reaching and range from reputational damage to extreme governmental surveillance.

Policymakers are in a never-ending rat race with technological innovation, and all the while IoT is becoming more widespread and affordable. The European Union's (EU's) General Data Protection Regulation (GDPR) empowers data subjects to have more visibility and control over their data and exercise their rights over it, which has significant implications for businesses and organisations large and small. Not only will the GDPR increase transparency and trust between data owners and subjects, but it will also encourage data controllers to think more closely about the legitimate business purpose of collecting the data and sharing it with third parties. It is only a matter of time before other nations adopt a similar standard for data protection.

In the future, individuals will need to take more ownership of their own data and understand what they are agreeing to in user agreements and privacy policies. This requires technology companies to be more transparent and educate their data subjects in plain language. Although disruption is key to IoT innovation and has potential far-reaching benefits including improved health and well-being and improved traffic flow, privacy by design is critical in ensuring the trust of consumers and avoiding potential personal or professional risks.

3.2 Data is at the Core of Privacy and IoT

3.2.1 What is Privacy?

Privacy is a fundamental human right recognized in the UN Declaration of Human Rights, the International Covenant on Civil and Political Rights and in many other international and regional treaties. Privacy underpins human dignity and other key values such as freedom of association and freedom of speech [3].

If a stranger asked you about your health and well-being, what you last bought on Amazon, and followed you around town, that might feel like an invasion of privacy. However, social media and apps constantly monitor and share your preferences in order to target your profile, and fitness tracking apps, such as Strava and Nike Run Club, monitor your location. In the new digital age, as the separation of online and offline is diminishing, it is hard to feel truly alone or private. One must question if consumers really want privacy. Digital natives have grown up with a wealth of information at their fingertips, and with virtual friendships rather than physical ones. According to Childwise, the average time spent in front of a screen, doing anything from playing games on a tablet to watching TV, increased from 3 h in 1995 to 6½ h in 2015 [4]. During that period of technological advancement, privacy was not at the forefront of people's minds. Consumers tend to want the newest gadget; a study from Barclays found that 62% of consumers would upgrade their smartphone in the next year [5]. Privacy in the context of IoT is a relatively new topic, but one that will become more pertinent as consumer adoption of IoT devices grows.

3.2.2 IoT and Data

Each IoT device has a unique ID and Internet Protocol (IP) address. Sensors in devices monitor the environment and humans in that environment, be it voice, temperature or blood sugar level. This sensor data is captured in microchips in the devices, which communicate to other devices on a network using Radio frequency identification (RFID) and Near-field communication (NFC). The data is then subjected to analytics to drive insights and actions. Actions could range from recommending a product to a consumer to increasing insulin levels in the bloodstream. Data is key for IoT, and can be broadly categorised into two buckets: personal data (including sensitive personal data), and non-personal data, which includes both proprietary data belonging to businesses, and open data such as traffic data. Companies that collect IoT data often harness it in data lakes or data fabrics, normally not making full use of the data and storing it indefinitely. Machine-learning algorithms are then applied to the big data in order to profile the users, drive insights and make recommendations. This can benefit consumers and corporations alike, increasing convenience for purchasing decisions, or creating medical recommendations. For example, Google revealed a way of using machine learning to personalise search ads to target consumers in 2018 [6].

IoT can itself be separated into two main buckets: the Internet of Humans, and the Internet of Machines. In the Internet of Humans, consumers freely share data with IoT devices to reap the benefits, often oblivious to the downstream implications of their data usage. Having a vast repository of personal data also allows companies to assign attributes to personas. Based on these personas, companies can then target individuals with information or advertising. IoT devices collect not only consumer data, but also employee data and proprietary business data, which is part of the Internet of Machines. The smart industry, or Industry 4.0, involves machine to machine (M2M) IoT device communications. Proprietary data captured from IoT devices could be highly confidential and contain intellectual property. Where confidential information is shared by organisations via an IoT network, this data could also be the subject of tampering or hacking; for example, the breach of this data could compromise a competitive advantage or give a competitor a head start in the race for a patent approval.

3.2.3 Data Considerations

A key challenge with the process of gaining insights from data is data quality. Companies often collect data without verification or deduplication (removing duplicate data entries). As such, data governance is key to ensuring one source of truth when it comes to personal data. Once this data is captured, much of the data is often left untouched, leading us to the question of why is it captured in the first place. According to a McKinsey Global Institute report, less than 1% of IoT data is currently used [7]. Organisations need to be more mindful about the personal data they capture on individuals in two ways. Firstly, they should only capture the data needed, and be selective over where this data is transferred, preventing unnecessary future complications. Secondly, if the data is no longer useful, then it should be deleted unless it is the subject of a litigation hold. If data is shared with multiple third parties, it is harder for the data to be deleted should the consumer exercise their right to request it. To make the deletion process easier, the data controller should map where the data flows. Transparency between data owner and data controller and data subject is key for IoT uptake.

3.3   Use Cases of Privacy and IoT

The Panopticon can be used as a metaphor for Privacy in the context of IoT. Jeremy Bentham's Panopticon is a building where a watchman in a tower watches the inhabitants in cells who do not know whether or not they are being watched at any time. The French philosopher Michel Foucault used it as an example of asymmetrical surveillance. "He is seen, but he does not see; he is an object of information, never a subject in communication" [8]. However, whereas in the Panopticon the inhabitants are aware of the possibility of being watched, in the digital age, there is no watchtower looming over us when we say "Alexa, switch off the lights". In this situation, the providers of IoT devices are the watchman, and the IoT devices are the watchtower. A stark parallel can thus be drawn between the Panopticon and IoT. Rather than our physical bodies being watched by a watchman, our actions and decisions are being captured as data not only by the government, but by corporations who create personas according to the data. At the extreme end of the spectrum, we are living in a surveillance state, or a digital panopticon (Fig. 3.1).

Time	7 am	7:30 am	8 am	9 am to 5 pm	6 pm	24/7
Context	Switch on the kettle	Morning run	Drive to work	At work	Evening in	Always-on
IoT Device	Alexa	Fitbit	CCTV	Implanted chips	Smart TV	Pacemaker
Sensor	Voice	Heart rate	License number plate	Location	Voice	Heart rate
Output	Kettle	Data	Crime monitoring	Increased efficiency	Ease of use	Healthcare

Fig. 3.1 A day in the life of IoT

3.3.1   Consumer Industry

In the consumer industry, corporations use Data Management Platforms to trace clicks from any devices, thus following a consumer journey, and can target with advertisements based on one's persona and device to ensure their product is at the forefront of the mind and easy to purchase. This becomes even more pertinent when devices such as Alexa, Amazon's cloud-based virtual assistant, are added to the mix. Rather than physically tapping a touchscreen, what if Alexa hears your conversations and recommends products to you? What if a wearable health device detects the symptoms of a disease and tells your phone to advertise the treatment for it on your Instagram, or on a loved one's Facebook? Although both potential situations may benefit the consumer in terms of convenience, there must be a balance between ease of purchase and personal privacy. The impact of a lack of privacy can affect more than just the end user. The Strava fitness tracker app uses GPS tracking to calculate distance travelled and elevation gained. Although these routes are not public, the app also provides a publicly available heat map that shows the exercise routes of all users. The users are all anonymised, but still the heat map inadvertently put lives at risk by sharing the routes of soldiers in secret military bases [9].

3.3.2   IoT for Healthcare

In the healthcare sector, personal medical devices are hailed as putting the control of healthcare in the hands of the patients. The role reversal of patients becoming teachers and doctors students is transforming the healthcare industry. Examples of this role reversal in action range from simple iPhone apps such as Ada Health which use artificial intelligence to analyse your symptoms and suggest disease states and treatments, to wearable devices which monitor your health remotely, providing real time feedback and potentially decreasing medical costs and the need for hospital beds. The benefit of having all of this data about yourself is also a risk. Whereas in the past, you would go to your doctor to have a test taken, which would be recorded on paper or a local computer (mainframe), these days, many of the solutions rely on Software as a Service (SaaS) or Platform as a Service (PaaS) solutions. In order to analyse the data and drive recommendations, the data captured from the device is sent to the "Cloud" where it is analysed. The output is then pushed back to the device. The risk here stems not only from the likelihood of the network of the IoT being hacked, but also by it being manipulated or stolen. For example, if a hacker is able to access the cloud where your medical information is held, they can one, hold it to ransom, as happened as part of the global cyberattack by WannaCry in 2017 [10], two, manipulate it to alter the dosage of drugs to potential toxic levels, three, sell it to unethical third parties and research companies, or four, use it for identity theft. These issues can all stem from a lack of privacy and security standards.

In April 2018, some 350,000 patients with a cardiac defibrillator were told that their devices were vulnerable to cybersecurity attacks [11]. The devices, manufactured by St Jude's Medical (which was acquired by Abbott Laboratories in January 2017), needed to have a firmware update, which required patients to visit their healthcare providers to get an upgrade. Prior to this, St Jude's was embroiled in another scandal in which the IoT company MedSec found vulnerabilities in the devices which could be life-threatening if hacked. St Jude's subsequently applied updates to the Merlin remote monitoring system, which is connected to cardiac devices [12]. Merlin.net is a patient care network, which works together with Merlin@home, a transmitter which monitors the device and sends the information to the doctor, reducing hospital visits. Many of the 350,000 patients are likely to be elderly, and may not understand the significance of malware patches and the risks associated with taking no action. Thus, it is the responsibility of the device manufacturers to clearly state the risks associated and effectively communicate this to the patients and doctors.

3.3.3 Smart Homes

IoT extends further than just wearable devices. Smart homes and offices have been hailed as a way to increase energy efficiency thanks to smart meters and sensors which detect your presence and alter the thermostat or lighting accordingly. An example of this is the remotely operated smart thermostat developed by Nest Labs, which uses machine learning on data captured from sensors to track people's schedules and adapt the temperature of their environment accordingly to conserve energy. Nest Labs is a smart automation company founded by Apple engineers and acquired by Google in 2014. In May of the same year, activists from the German group Peng! Collective pretended to be Google representatives to unveil a new site Google Nest at the Re:public tech conference. Google Nest used Google's iconography to poke fun at its privacy policies, and offered four products that were unnervingly believable [13]. All four products captured personal information for different purposes:

- Google Trust "The more Google products you use, the higher your insurance payout will be in case of data misuse through secret service or private criminals."
- Google Hug: "Google Hug helps you find others nearby who have the same needs you do."
- Google Bee: "Introducing the first personal drone. Google Bee watches over your house and family when you are away".
- Google Bye, a memorial site: "Each time you use a Google service, like downloading an app from the Play Store or watching a video on YouTube, you tell us a little bit more about yourself. Why leave that valuable information with us when you can share it with others?"

The site was so convincing that people emailed it asking for more information about the features offered.

The concerns associated with lack of privacy can be manipulated not only by individuals and organisations, but also by governments. In 2017, Wikileaks revealed that the US Central Intelligence Agency (CIA) had tapped into smart devices, including Samsung's smart TVs. In leaked notes from a "Weeping Angel" workshop the CIA conducted with the UK's Security Service (MI5), detailed instructions were given about software which enabled a Fake-Off mode. This allowed the CIA to listen in to conversations and send them to a CIA server, even when the TV is apparently switched off, and included "Suppress LEDs" to make the Fake-Off mode look better [14]. Michael Hayden, an ex-CIA director responded to this by trying to reassure citizens: "These tools would not be used against an American", and "There are bad people in the world that have Samsung TVs, too" [15]. This begs the questions: are all Americans good? What defines an American?

3.3.4 Smart Offices

Security is an advantage that comes with smart homes and offices; the McKinsey Global Institute (MGI) states that "by using digital security cameras with advanced image-processing capabilities, operators of office buildings can monitor activity throughout their properties without requiring guards to patrol or continuously monitor video feeds" [7]. In workplaces, IoT can track workers in real time to warn them when they are entering an area of concern, and some modern employers are also using IoT to increase efficiency by tracking employees. In Sweden, the technologically advanced company Epicenter put RFID chips under the skin of staff to enable them to gain access to the building or use the photocopier [16]. Although this is improving safety and efficiency, this is a prime example of forsaking privacy for extreme convenience.

3.3.5 Smart Cities and Crime

At a higher level, smart cities will offer improved transportation efficiency. By putting IoT devices on public transport vehicles, the location of each vehicle can be tracked at any point in time, telling you how far away your bus is. By tapping your pass on the bus, your location can also be mapped to that bus. Some governments are also tracking vehicle license plates using cameras to catch drivers who are speeding. In these situations, there is no process in place to gather consent from the drivers.

Privacy becomes more muddied in the context of communities. The Baltimore Police Department deployed Cessna airplanes with surveillance cameras to fly over the streets of Baltimore with the purpose of monitoring crime, to the ignorance of innocent civilians [17]. As such, the move to smart homes and smart cities raises concerns about transparency. In Toronto, Canada, Google's parent company Alphabet is pioneering Sidewalk Labs in Quayside on the banks of Lake Ontario [18]. Canadian Prime Minister Justin Trudeau coined it as a "thriving hub for innovation, [to] create the good, well-paying jobs that Canadians need." Sidewalk Labs boasts automatic waste collection, autonomous vehicles and connected transit systems. Google called it "a distributed network of sensors to collect real-time data about the surrounding environment." How will this work in terms of privacy? With facial recognition on the streets, thoughts will be the only thing which will remain private, especially since eye-tracking software can already detect an individual's mood solely based on facial movements. Will a privacy notice be served to the residents every time a new IoT device is deployed?

3.3.6 Protecting Privacy

A new trend of Edge Analytics has the potential to curb many of these risks described. Edge analytics relies on the local computing power of the IoT device and other devices closer to the edge of the network, rather than that of the cloud network. If analytics and decision making can be done locally, then this mitigates the risks associated with the cloud being hacked. In addition, the local hosting of data leads to increased data privacy, a decreased risk of hacking, and increased data ownership by the consumer rather than the supplier. Of course, there is still the possibility of the device itself being breached; however, this is less favourable for hackers than breaching a single point of security which touches multiple users. Blockchain also has the potential to increase privacy in IoT. Rather than sending data to a central corporate-owned cloud, what if the data was decentralised? This way, only metadata containing no personal information is transferred between systems.

The importance of privacy can be found in many use cases for IoT, as discussed. The movements and decisions humans make are datafied by IoT, and IoT can subsequently dictate our next actions. Consumers can reap benefits from IoT in their daily lives. For example, Alexa listens to commands which can control the lighting in your house to go the time you wake up in the morning. Alexa can also connect to smart automated home systems such as your smart kettle. As a McKinsey Global Institute paper on IoT states, the opportunities of IoT in the home can give consumers time back due to the automation of domestic chores, and increase energy efficiency [7]. The variety of use cases requires privacy standards which can be applied to all these situations. It is time suppliers begin to share best practice in ensuring the privacy of users is respected. Proportionality is required to ensure technological innovation can proceed. One way in which this can be done is by categorising personal data depending on its sensitivity, and putting in suitable controls based on that category of data.

3.4 The impacts of Privacy (or Lack Thereof) on Our Daily Lives

3.4.1 Consumer Demand for IoT

Philip N Howard writes in his book *Pax Technica: How the Internet of Things May Set Us Free or Lock Us Up* that the Internet of Things will lead to a new political age he calls the "Pax Technica", where the government and technology companies have hegemony over citizens [19]. However, the societal impact of privacy is yet to emerge in full force. This is due to the relatively slow uptake of smart devices by consumers. Only 6% of American households have a smart-home device, including internet-connected appliances, home-monitoring systems, speakers or lighting, according to Frank Gillett of Forrester, a research firm. A survey conducted by PricewaterhouseCoopers found that 72% of people did not expect to adopt smart-home technology over the next 2–5 years. Forrester predicts that growth will be slow, with around 16% of American households with a smart-home device in 2021 [20]. The reason for this can be boiled down to the cost of the devices combined with the lack of necessity. A Samsung fridge costs $5000, and do you really need to check what is left in the fridge without opening the door? The Economist also attributes the slow uptake to the low turnover of smart-home devices. The smart-device ecosystem is still disparate, and in order for it to be successful, more interoperability on open platforms is needed.

3.4.2 Consumer Privacy

The Amazon Echo is an example of a smart device that is performing well. The success of the Amazon Echo is due to the affordability of the Echo device and its commerce revenue stream. The investment firm Mizuho predicts that by 2020, $4 billion will come from the sales of the Echo devices, and a whopping $7 billion Echo related revenue generated from the commerce on Amazon.com [21]. Amazon is an example of a company which excels at customer engagement and retainment, and prides itself on its customer obsession. When a consumer searches for a product on Amazon, he/she receives email reminders from Amazon, and advertisements on other websites, pushing the product to the forefront of the consumer's mind. Although Amazon states that Echo devices do not listen in to all conversations, and only listen when the word Alexa is used, it emerged in May 2018 that a private conversation between a lady called Danielle and her husband was recorded and sent to a random contact, without consent [22]. Fortunately, the conversation was about hardwood flooring. Nevertheless, Danielle called it "a total privacy invasion. Immediately, I said, 'I'm never plugging that device in again because I can't trust it.'" Although in this situation, Amazon claimed it was an extremely rare occurrence due to Alexa mishearing trigger words rather than listening to all conversations,

Amazon has filed patent applications for "voice-sniffing" algorithms which constantly listen for words such as "love" and if they "bought" something, analyse the speech, adding it to a database [23]. Following this, the patent application even goes on to propose a way to serve targeted advertisements to not only the consumers themselves but also their friends and family members.

Amazon's motivations outlined above raise many questions about privacy, which the more vulnerable members of the population are not asking. In particular, the elderly, who grew up during a time before the internet was invented, may not appreciate how to protect their privacy online. There is a whole host of new terminology constantly being birthed, often by millennials, which can be hard to keep up with and understand. "We use cookies to provide you with the best experience on our site" is a common phrase found on most websites. Many people do not know that cookies capture site name and a unique user ID, meaning that when you visit the website again, your PC tells the site, which can then personalise the content based on previous visits. Some cookies are more sophisticated. They might record how long you spend on each page on a site, what links you click, even your preferences for page layouts and colour schemes. They can even be used to store data on what is in your "shopping cart". This means that you could receive targeted offers based on your browsing history. The new EU ePrivacy regulation, which is yet to be implemented, explicitly points to IoT in the proposal "the principle of confidentiality which is enshrined in the Regulation should also apply to the transmission of machine-to-machine communications [24]." The ePrivacy Regulation aims to simplify rules governing cookies, and gives individuals more rights around electronic communications.

3.4.3   Privacy Education

The lack of general understanding around technology in society has led to many consumers being oblivious to the lack of data privacy, and at the other extreme, technophobes. As such, education is key in ensuring a balanced approach to the use of technology in our daily lives. With attention spans decreasing, we are unlikely to read through the privacy notice on websites, or change our privacy settings on Facebook. Often, it is only when a breach of privacy occurs, that action to protect privacy is taken. More needs to be done by regulators and technology companies to communicate how to protect privacy in a way that is intelligible by society. The way that this is done is also crucial. Rather than having pages of long documentation, regulators and technology firms should use their abundant innovative resources and invest in new ways to capture the attention of consumers using modern communication channels, such as infographics, videos and virtual reality. Facebook has done this by sharing bite-size clips about their privacy settings, and advertising on billboards; however, it may be too little, too late for them. Only once individuals truly understand how their data is being processed, will they take ownership and accountability over it. If identity theft and catfishing, which is where someone pretends to be another person online to date someone, can occur from data captured on personal computers (PCs), the risk becomes even greater if hackers can access smart devices.

3.4.4 Societal Impacts and Discrimination

"Nudge theory" is a behavioural psychology concept explained in the book *Nudge: Improving Decisions About Health, Wealth, and Happiness* by Richard Thaler and Cass Sunstein and involves adapting the environment to trigger cognitive processes [25]. Wearable devices such as Fitbits and Apple Watches claim to help consumers reach their health and fitness goals. One major motivating factor for individuals to exercise more is the community aspect. Some people do not go for a run unless they have Strava switched on, so they can share their exercise with others. This is an example of positive data sharing. The gamification of fitness using apps brings out the competitive nature of individuals, such as various step challenges where different teams compete to rack up the most steps.

Conversely, the rise in technology has also been blamed as a major cause of the incidence of increasing mental health issues. The herd nature of social media can give strength to those who have the same opinions, but also alienate individuals with differing opinions, and this can be perpetuated by the community aspect of wearable devices. Having your data out in the open leaves it open to judgement and profiling. However, some digital natives are defying the norm and generating new ways to succeed in the modern world, and there are ways in which IoT can combat these issues. In 2017, school children came together in the UK to develop ways to do this. Innovative ideas include a smart wearable band "Breath Watch: a wristband that monitors the symptoms of a panic attack and provides calming down techniques via a mobile phone or tablet [26]." Interestingly, Public Health Canada takes advantage of the lack of online privacy by combing social media posts for indicators of mental health and likelihood of suicide [27].

The lack of privacy associated with sensitive personal data such as medical data captured from wearable devices can have a detrimental impact on insurance and employability. Even non-sensitive personal data can lead to discrimination. Lawyers in the USA have used big data and machine learning to calculate the probability of criminals reoffending, and police in the UK have used technology for "predictive crime mapping" [28, 29]. Although these examples are in the best interests of ensuring citizens' safety, one of the biggest markers of likelihood of crime and reoffending was location and race. Because the data used to train the algorithm was inherently biased, the algorithm learned and reinforced racial bias, effectively discriminating against certain races. Currently, the algorithms use crime type, location, date and time; however, the potential of adding data captured from IoT devices to the mix could raise more controversial issues. In these hypothetical circumstances, it is important to distinguish between correlation and causation.

3.4.5   Governmental Monitoring

The lack of privacy can be manipulated by governments, as well as individuals, and organisations. In 2013, Edward Snowden revealed the state of extreme surveillance engineered by the US government, alongside the UK, Australia and Canada [30]. The National Security Agency (NSA) was able to access Google and Yahoo accounts as well as phone calls, and thus snoop on citizens' personal and professional lives. Ironically however, the documents which showed the lack of privacy, also inadvertently breached the privacy of agents, forcing them to move location for their security. In 2016, The Guardian released an article where James Clapper, the US director of national intelligence, confessed that "in the future, intelligence services might use [IoT] for identification, surveillance, monitoring, location tracking, and targeting for recruitment, or to gain access to networks or user credentials" [31]. Despite all of this, there is no single federal law regulating collection and use of personal data in the USA as of 2018.

3.5   Privacy Regulations and Their Implications for Organisations and Individuals

The Cambridge Analytica data scandal, among others, have highlighted the importance of data protection in the digital age where technology is advancing at a rapid pace and data is becoming the new oil, both by fueling the economy of the future and increasing in value. Websites, including the big tech giants, feed off consumer data to generate revenue from targeted advertising, and this often benefits consumers, who can get attractive deals relevant to them. The risk lies when the data is misused or breached, and consumers are at the mercy of the data controller.

3.5.1   The EU General Data Protection Regulation

The EU General Data Protection Regulation (GDPR) puts control back in the hands of consumers. It came into effect on 25 May 2018, and aims to harmonise data protection laws across the EU and give regulators stronger enforcement powers. The same rules apply to international companies who process EU personal data, and this led to the EU-wide shutdown of websites such as the LA Times who were not ready for the deadline [32]. On the plus side, others are now using the GDPR as best practice, with the Salesforce CEO Marc Benioff saying that the USA should follow suit and implement a similar data privacy law [33]. The GDPR applies to IoT, as it mentions RFID tags in its list of online identifiers. There are two main ways the GDPR is empowering consumers: through increased transparency, and bolstered consumer rights. The tirade of emails asking us to keep in touch have been a direct result of the GDPR requirement to serve a privacy notice and ask for explicit, opt-in consent where legally required. This requirement is a great start in encouraging companies to consider and clearly communicate what data they are capturing (data minimisation), what they are using the data for, and how long they are keeping the data (retention policies). The burden of increased scrutiny and tighter regulations has meant that organisations tend to capture less sensitive personal information, unless it is crucial. GDPR has brought to light the wealth of archived and unused data, and led to a conscious decision to either contact the data subject to keep storing the data, or undergo a spring clean of the unused data.

The GDPR is a great first step in regulating data processors, but it is important to be realistic about the benefits. An example of this is Article 13 of the GDPR "Information to be provided where personal data are collected from the data subject" which refers to a privacy notice [34]. According to Emily Taylor, associate fellow of Chatham House, it would take 250 h per year to read all of the privacy notices of services from start to end [35]. Consumers tend to choose convenience over control. When you are in a foreign location, it is easy to reach for your phone and see what restaurants are nearby, and how to get there. As such, it is important for companies to show that they value their consumers' data rights by making the privacy notice more easily accessible, and clearly stating which second and third parties process the data.

A common misunderstanding about the GDPR is that consent is required for processing data. Consent is only required where there is no legitimate business purpose for processing the data, such as for direct marketing purposes. This becomes tricky in some situations, such as a retail context where a store wants to send you targeted vouchers in store based on your previous shopping habits. An example of the how the GDPR strengthens individual rights is shown by the new Facebook privacy settings, which now give consumers the ability to see how their data is processed, or to be forgotten completely. The right to be forgotten is a major challenge for IoT devices. Although the GDPR aims to harmonise data protection rules across the EU and European Economic Area (EEA), it is important to consider other regulations that may overrule it. For example, clinical trials data, which could be captured from wearable health monitors, must be retained for 20 years after completion of the study, according to the Medical Research Council (MRC) in the UK [36].

3.6 Conclusion

The Helsinki Privacy Experiment of 2012 used 10 volunteer homes to investigate the "Long-term Effects of Ubiquitous Surveillance in the Home" over a 6-month period [37]. Sensors were implemented to track network traffic, personal computers, smartphones, cameras and payment cards. Interestingly, they found that despite initial aversion to monitoring, people became accustomed to the lack of privacy. One participant said that once he had accidentally been seen naked by the camera, his threshold for privacy was lowered. Outside of the confines of the experiment, societies are becoming normalised to the lack of privacy as the desire for convenience and exciting new technology outweighs the risks of lack of privacy. Companies should take accountability for educating consumers on the rights they have on their personal data, and all corporations should be transparent about the way they process personal data, whether they are under control of the GDPR or not. Valuing the privacy of individuals should be considered by organisations as a competitive advantage, as it increases the trust between consumers and corporations, and between citizens and governments.

In this age of surveillance capitalism where we are moving ever closer to a digital panopticon, there has been a gap in privacy regulation and technological innovation, which has led to regulators scrambling to keep up. Nevertheless, the uptake of IoT devices has been slower than expected and this gives regulators an opportunity to leapfrog and cement their standards before IoT devices are deployed in even greater numbers around the world.

One must look at both sides and not forget the advantages of collecting vast troves of data from IoT devices. Data can be used for good—from personalising an individual's user experience to managing crime and monitoring health and wellbeing, decreasing the strain on health providers. IoT providers will increasingly need to communicate the value consumers gain when providing data. Proportionality is essential in ensuring a balance between technological innovation and privacy, a fundamental human right. The rapidly changing environment in which we live requires individuals to take ownership of their privacy, championed by governments and organisations. With the blurring of lines between online and offline, humans and artificial intelligence, we must not lose sight of what makes us human.

References

1. Royal Society (2017) The Internet of Things: opportunities and threats. Conference report. https://royalsociety.org/~/media/events/2017/10/tof-iot/iot-conference%20report-final.pdf. Accessed 14 Apr 2018
2. Cadwalladr C, Graham-Harrison E (2018) Revealed: 50 million Facebook profiles harvested for Cambridge Analytica in major data breach. https://www.theguardian.com/news/2018/mar/17/cambridge-analytica-facebook-influence-us-election. Accessed 14 Apr 2018
3. Banisar D, Davies S. Privacy and human rights. An International Survey of Privacy Laws and Practice. http://gilc.org/privacy/survey/intro.html. Accessed 14 Apr 2018
4. Wakefield J (2005) Children spend six hours or more a day on screens. https://www.bbc.co.uk/news/technology-32067158. Accessed 20 May 2018
5. Potuck M (2017) Barclays: 35% of consumers intending to buy an iPhone going with X. https://9to5mac.com/2017/12/04/iphone-x-barclays/. Accessed 20 May 2018
6. Kalischko T (2018) Google Marketing Live 2018 – Age of Assistance. https://smarter-ecommerce.com/blog/en/adwords-automation/google-text-ads/google-marketing-live-2018/. Accessed 27 Oct 2018
7. Manyika J, Chui M, Bisson P, Woetzel J, Dobbs R, Bughin J, Aharon D (2015) Unlocking the potential of the Internet of Things. https://www.mckinsey.com/business-functions/digital-mckinsey/our-insights/the-internet-of-things-the-value-of-digitizing-the-physical-world. Accessed 12 May 2018

8. Michel F (1977) Discipline and punish: the birth of the prison. Pantheon Books, New York. Print

9. Hern A (2018) Fitness tracking app Strava gives away location of secret US army bases. https://www.theguardian.com/world/2018/jan/28/fitness-tracking-app-gives-away-location-of-secret-us-army-bases. Accessed 27 May 2018

10. BBC (2017) NHS cyber-attack: GPs and hospitals hit by ransomware. https://www.bbc.co.uk/news/health-39899646. Accessed 14 Jul 2018

11. Vaas L (2018) 350,000 cardiac devices need a security patch. https://nakedsecurity.sophos.com/2018/05/04/half-a-million-pacemakers-need-a-security-patch/. Accessed 14 Jul 2018

12. US FDA (2018) Battery Performance alert and cybersecurity firmware updates for certain Abbott (formerly St. Jude Medical) implantable cardiac devices: FDA safety communication. https://www.fda.gov/MedicalDevices/Safety/AlertsandNotices/ucm604706.htm. Accessed 14 Jul 2018

13. Hullinger J (2014) How activists fooled the internet with these convincing new Google nest products. https://www.fastcompany.com/3030418/how-activists-fooled-the-internet-with-these-convincing-new-google-nest-products. Accessed 14 Jul 2018

14. Frenkel S (2017) US intelligence officials: latest WikiLeaks drop "Worse Than Snowden" Docs. https://www.buzzfeed.com/sheerafrenkel/us-intelligence-officials-latest-wikileaks-drop-worse-than-s?utm_term=.al62x3vpJ#.we6e14YzE. Accessed 14 Jul 2018

15. Nguyen N (2017) If you have a smart TV, take a closer look at your privacy settings. https://www.cnbc.com/2017/03/09/if-you-have-a-smart-tv-take-a-closer-look-at-your-privacy-settings.html. Accessed 14 Jul 2018

16. Cellan-Jones R (2017) Office puts chips under staff's skin. https://www.bbc.co.uk/news/technology-31042477. Accessed 7 Jul 2018

17. Reel M (2016) Secret cameras record Baltimore's every move from above. https://www.bloomberg.com/features/2016-baltimore-secret-surveillance/. Accessed 7 Jul 2018

18. Hemmadi M (2017) Google's Sidewalk Labs project in Toronto is already creating jobs—in New York. https://www.macleans.ca/news/canada/googles-sidewalk-labs-project-in-toronto-is-already-creating-jobs-in-new-york/. Accessed 7 Jul 2018

19. Powles J (2015) Internet of things: the greatest mass surveillance infrastructure ever? https://www.theguardian.com/technology/2015/jul/15/internet-of-things-mass-surveillance. Accessed 7 Jul 2018

20. The Economist (2016) Where the smart is. https://www.economist.com/business/2016/06/11/where-the-smart-is. Accessed 7 Jul 2018

21. Kim E (2016) Amazon's Echo and Alexa could add $11 billion in revenue by 2020. http://uk.businessinsider.com/amazon-echo-alexa-add-11-billion-in-revenue-by-2020-2016-9?r=US&IR=T. Accessed 7 Jul 2018

22. Horcher G (2018) Woman says her Amazon device recorded private conversation, sent it out to random contact. https://www.kiro7.com/news/local/woman-says-her-amazon-device-recorded-private-conversation-sent-it-out-to-random-contact/755507974. Accessed 28 Jul 2018

23. US Patent and Trademark Office (2018) http://appft.uspto.gov/netacgi/nph-Parser?Sect1=PTO2&Sect2=HITOFF&p=1&u=%2Fnetahtml%2FPTO%2Fsearch-bool.html&r=1&f=G&l=50&col=AND&d=PG01&s1=amazon.AANM.&s2=conversational&OS=AANM/amazon+AND+conversational&RS=AANM/amazon+AND+conversational. Accessed 28 Jul 2018

24. i-Scoop (2018) The new EU ePrivacy regulation: what you need to know. https://www.i-scoop.eu/gdpr/eu-eprivacy-regulation/. Accessed 28 Jul 2018

25. Kobie N (2015) Nudge theory: when your smart gadgets nag you. https://www.theguardian.com/technology/2015/aug/07/nudge-theory-smart-gadgets-silicon-valley. Accessed 14 Jul 2018

26. Nesta (2017) School children use the Internet of Things to tackle mental health issues. https://www.nesta.org.uk/news/school-children-use-the-internet-of-things-to-tackle-mental-health-issues/. Accessed 28 Jul 2018

27. The Canadian Press (2018) Federal health agency to mine social media for study on suicide trends, risk factors. https://www.theglobeandmail.com/life/article-federal-health-agency-to-mine-social-media-for-study-on-suicide-trends/. Accessed 28 Jul 2018
28. Buranyi S (2017) Rise of the racist robots – how AI is learning all our worst impulses. https://www.theguardian.com/inequality/2017/aug/08/rise-of-the-racist-robots-how-ai-is-learning-all-our-worst-impulses. Accessed 28 Jul 2018
29. Dearden L (2017) How technology is allowing police to predict where and when crime will happen. https://www.independent.co.uk/news/uk/home-news/police-big-data-technology-predict-crime-hotspot-mapping-rusi-report-research-minority-report-a7963706.html. Accessed 28 Jul 2018
30. Ball J, Borger J, Greenwald G (2013) Revealed: how US and UK spy agencies defeat internet privacy and security. https://www.theguardian.com/world/2013/sep/05/nsa-gchq-encryption-codes-security. Accessed 28 Jul 2018
31. Ackerman S, Thielman S (2016) US intelligence chief: we might use the internet of things to spy on you. https://www.theguardian.com/technology/2016/feb/09/internet-of-things-smart-home-devices-government-surveillance-james-clapper. Accessed 28 Jul 2018
32. Hern A, Belam M (2018) LA Times among US-based news sites blocking EU users due to GDPR. https://www.theguardian.com/technology/2018/may/25/gdpr-us-based-news-web-sites-eu-internet-users-la-times. Accessed 28 Jul 2018
33. Farber D (2018) Salesforce CEO Marc Benioff Calls for National Privacy Law. https://www.salesforce.com/company/news-press/stories/2018/5/051618/. Accessed 28 Jul 2018
34. Intersoft Consulting (2018) https://gdpr-info.eu/art-13-gdpr/. Accessed 28 Jul
35. The Royal Society (2017) [starting at 30:00] https://www.youtube.com/watch?v=C_QXOHX5xSA&list=PLg7f-TkW11iWmGlFJ9-IkIffIcwRt9s74&index=3&t=0s. Accessed 28 Jul 2018
36. Medical Research Council (2017) MRC Regulatory Support Centre: Retention framework for research data and records. https://mrc.ukri.org/documents/pdf/retention-framework-for-research-data-and-records/. Accessed 28 July
37. Oulasvirta A, Pihlajamaa A, Perkiö J, Ray D, Vähäkangas T, Hasu T, Vainio N, Myllymäki P (2012) Long-term effects of ubiquitous surveillance in the home. Proceedings of the 2012 ACM conference on ubiquitous computing. UbiComp' 12:41–50

Chapter 4
A Business Framework for Evaluating Trust in IoT Technology

Fen Zhao and Britt Danneman

4.1   Uber: A Case Study in Trust

Few of us stop to consider how much stepping into a taxi or hired car makes us vulnerable. As a passenger, we do not have control over where the car drives, how safely it drives, or where it stops. With a traditional taxi, at least, this sense of vulnerability is reduced by social norms developed and reinforced over the course of decades—your parents and possibly your grandparents hailed cabs, too, so you do not even question doing it. The existence of formal credentialing, in the form of the taxi medallion, lessens feelings of vulnerability still further.

Ride-hailing apps like Uber, which marry real-world car services with the internet, dramatically alter this dynamic. Instead of a professional cab driver, each car is driven by a random stranger who may have only been given a cursory background check.

The idea of getting in a stranger's car used to activate an instinctive fear response in most people. It went against one of the first safety rules most of us learned as children. However, ride-hailing apps add a technological "trust layer" that enables users to overcome that fear even in the absence of social norms around the service. The Uber app collects a review from each passenger after their ride, making it easy to crowdsource trust and also flag and remove bad drivers. Uber also tracks all cars by GPS, so every ride is traceable. This makes it easy for users to evaluate performance history, a key condition for cultivating trust online, according to Friedman, Kahn Jr., and Howe [1].

These trust bridges have helped users make a trust leap [2]. As of June 2018, according to eMarketer, more than 24% of Americans had used a ride-sharing app, up from only 5% 4 years before [3]. In under a decade, these innovations have powered

F. Zhao (✉) · B. Danneman
Alpha Edison, Los Angeles, CA, USA
e-mail: fen@alphaedison.com; britt@alphaedison.com

© Springer Nature Switzerland AG 2019
F. D. Hudson (ed.), *Women Securing the Future with TIPPSS for IoT*, Women in Engineering and Science, https://doi.org/10.1007/978-3-030-15705-0_4

Uber's ascent from a small startup to a company with more than $24 billion in funding and a market value estimated to be as high as $120 billion [4, 5].

Yet as with any IoT product, there is still a real question of how much users can trust Uber's service. That question is not just about the technology that underpins app-based ridesharing—it inevitably involves the reputation of Uber the company also. To trust any IoT product or service, users need to trust the people, process, and technology delivering that product or service as well.

Female riders especially may be concerned about Uber's reputation as a company that treats women poorly. A February 2017 blog post by a former Uber engineer detailing her experience of sexual harassment at the company—and the efforts of human resources (HR) to cover it up—caused an uproar that ended with the firing of Uber's founder as CEO and a major reorganization of Uber management [6]. Though Uber claims to be building a less toxic, more female-friendly workplace culture, there are signs that its problems are not quite over yet. As of April 2018, for example, at least 103 Uber drivers in the USA had been accused of sexually assaulting or abusing their passengers, according to CNN [7]. Some critics charge that Uber's policy of conducting its own background checks has created an environment that lets such behavior thrive.

None of this has to do with design features of Uber's app—but all of it might race through a potential rider's head as she decides whether to pull out her smartphone or hail a traditional taxicab. While ride-hailing apps are by no means value-neutral, the existence of a competitor (Lyft) that functions nearly identically but is trusted very differently exemplifies the importance of provider-user relationships in understanding trust in technology. The question of whether to trust ride-hailing technology is hard—perhaps impossible—to separate from the user's trust in the company that is the largest incumbent in the market.

4.1.1 Companies as Well as Technologies Build Trust

Existing frameworks for understanding user trust in technology are not always fully adequate for understanding the role of trust in companies in the development of that trust. Today the fields of human–computer interaction (HCI) and value sensitive design (VSD) focus on the suitability of technologies to the values of their users to explain the growth of trust [8]. An approach to technology design first developed at the University of Washington in the 1990s, VSD aims to control and shape the moral and political impact of technology on human lives by "account[ing] for human values in a principled and comprehensive manner throughout the design process" [9 p1]. While some critics have charged that VSD falls short of its goals—for example, by failing to define relevant "human values" clearly [10]—it remains an established approach to understanding how values like trust relate to technology.

While VSD researchers have recognized that "[p]eople trust people, not technology" [1 p36], it remains difficult to apply VSD frameworks to measure and interpret the impact of trust in a technology provider on user values around that technology. Most work focuses on user behaviors in the abstract or consists of empirical historical studies where, too often, attitudes around trust in a technology are convolved with trust in the single provider of that technology.

However, in most cases, the *technology provider* is a brand or company with its own unique history and relationship with the user that is the basis of its own unique trust relationship. That relationship can and should be treated separately from trust arising out of technology features. For example, as previously mentioned, users may have built a very different trust relationship with Uber than with Lyft, a competing ride-hailing service that has not been plagued by similar scandals.

In particular, IoT contains many opportunities for forging innovative trust relationships by refashioning or displacing older ways that trust was formed. It also opens possibilities for thinking about additional dimensions of trust, such as the role of the physical versus the digital world in building trust and in framing responses to trust violations. A more nuanced understanding of the interrelationship of trust in companies and trust in technologies is necessary to examine these facets. Similar to how VSD focuses on design choices for a technology, our approach focuses on design choices for a company, particularly the structure of its business models and brand value.

Understanding the texture of a user's trust relationship with a company begins with understanding its various elements. Behavioral scientists in management and marketing sciences frame users' trust in a company through the three pillars of *ability*, *integrity*, and *benevolence* [11, 12]. Does the company have the *ability* to fulfill its promises to the customer—for example, by successfully taking you from point A to point B? Does it have enough *integrity* to make morally correct decisions? And finally, does it have *benevolence* in the sense that it will put the customers' needs first? Events that call into question any of these perceptions may impact overall trust, so trust relationships may evolve over time.

This kind of trust becomes particularly important to consider when a company is a first mover within a technology area or has dominant market share. The trust relationship it establishes with users may determine the normative trust attitudes in society for that entire branch of technology. For example, Siegrist found that social trust in institutions that use genetic engineering or use genetically engineered products strongly influences users' perception of the riskiness of gene technology [13]. It is likely that users' trust relationships with Uber affect their overall trust in ride-hailing apps and associated technologies.

In the business context, understanding the evolution of trust over time is important for another key reason: It indicates opportunity to increase enterprise or brand value. For example, Lyft users are twice as likely to rate the company as "trustworthy" as they are other ride-hailing services [14]. This suggests that trust is a major source of value for Lyft's brand and a key reason it has been able to erode the monopoly of dominant player Uber. As venture capitalists who invest in early-stage companies, we, the authors, are particularly interested in how a deeper understanding of trust can inspire new business models and/or new products and services.

You may have noticed that the scope of our definition of IoT is broader than most others used in this book. Our work focuses on what could more accurately be called "cyberphysical systems," in which IOT is often a key component. (For instance, a ride-hailing app is part of a cyberphysical system that brings a physical car based on a digital hail.)

This broader category both encompasses and transcends the traditional definition of IoT, picturing it not just as an internet of things, but as an internet of services.

4.2   What Is Trust in a Business Context?

Trust is not a factor investors commonly take into account when estimating the value of an idea or company. That is in part because it is so difficult to measure. In this section, we lay out a qualitative understanding of what builds trust in a brand or company and a framework for arriving at a quantitative measurement of what that trust is worth. To quantify trust, we equate it with dollar value, the natural unit of measurement in business. Specifically, we look at fluctuations in the market value of a company as measured by its stock price.

As mentioned previously, we define the depth of users' trust in a company based on how the company measures up against three pillars: ability, integrity, and benevolence. As noted by McKnight, Cummings, and Chervany, most trust beliefs studied by behavioral researchers in marketing tend to cluster into these three categories. A fourth category, predictability, refers to the consistency of a company's behavior over time [11]. In this chapter, we treat this component separately as part of the "history" of a user's trust relationship with a brand.

Usually, ability, integrity, and benevolence all play into the trust relationship, and all three can play into the deepening of that relationship over time. In Table 4.1, we have arranged each element next to its definition in existing research. We have also given some examples of statements users could make about Uber that would indicate trust built on that element.

From this grid, it is easy to see that each of Uber's scandals touches on different elements of trust. Its HR issues with its own employees are mostly an integrity and benevolence issue from the perspective of the user; however, reports of sexual assault have a direct impact on ability, as they may cause users to doubt how safe riding with Uber will be.

You will notice that, as we move down the grid, the basis for users' trust in a company becomes increasingly abstract. It shifts from a trust based on expectation

Table 4.1 The Three Core Elements of Trust in Companies

	Definition	Example statements
Ability	Ability to do what the user needs	• Uber gets me from point A to point B quickly • Uber is safe and easy to use
Integrity	Honesty and promise-keeping	• Uber refunds me if something is wrong with my trip • Uber's prices are fair
Benevolence	Caring and motivation to act in the user's interests	• Uber treats its employees well • Uber supports female victims of sexual harassment and assault

of tangible, real-world results (ability) toward questions of values and intent (integrity and benevolence). This abstraction is one reason that the value of trust, or even its existence in a particular relationship, is difficult to quantify.

4.2.1 How Companies Build Trust with Customers

According to Friedman, Kahn Jr., and Howe, "we trust when we are vulnerable to harm from others yet believe these others would not harm us even though they could" [1 p34]. Vulnerability is a prerequisite for trust: For a trusting relationship to grow, one party or the other must put themself at risk [15].

Too often, it is the user who is forced into a position of vulnerability vis-a-vis the technology provider. For a company, showing trust—that is, making itself vulnerable—is one of the best ways to build trust. Some strong evidence for this exists in management literature. Prusak and Cohen note that "in companies that display trust, both toward employees and toward customers and suppliers, people are more likely to trust each other" [16]. When companies do not display such trust, the opposite happens. Prusak and Cohen cite this anecdote from Hewlett Packard founder David Packard's memoir "The HP Way," where he recounts his time working at a General Electric (GE) plant in the 1930s.

> GE was especially zealous about guarding its tool and parts bins to make sure employees didn't steal anything. Faced with this obvious display of distrust, many employees set out to prove it justified, walking off with tools or parts whenever they could. Eventually, GE tools and parts were scattered all around town, including the attic of the house in which a number of us were living.

By showing distrust and refusing to display vulnerability, GE accidentally fostered the very behavior it feared in its employees [16].

Vulnerability in the business context can take many forms. The most prevalent we see is around customer service or financial terms. Southwest Airlines, for example, makes itself financially vulnerable to customers with a policy it calls "transfarency," which means the company does not charge change or baggage fees. In the moment when the customer is most vulnerable (needs to change her flight or is told her bag is too big), Southwest takes the financial risk of flight change costs, and in doing so shows its benevolence and alignment with customers.

However, not all forms of vulnerability are created equal. Taking on financial risk may show benevolence, but it is still a benevolence rooted in commercial transactions—it does not take the customer relationship deeper. By contrast, companies that *commit publicly to values* or that *invest in and cede power to a community* pull users further down the trust funnel. For example, the online neo-bank Aspiration (an Alpha Edison portfolio company) publicly commits to values like sustainability. Not only does Aspiration offer sustainability-focused funds to investors, it is also transparent about how it manages its own investments; as the company's website puts it, "unlike the Big Banks, we do not use your deposits to fund big oil pipelines" [17]. By "letting go" and showing vulnerability in this way, companies like Aspiration are able to cultivate unprecedented levels of trust. In a forthcoming paper, we discuss the concept of trust levels for Aspiration and other companies in more depth [18].

Those trust relationships are not, however, stable over time. Events both positive and negative can change users' conceptions of a company's ability, integrity, and benevolence. There is no doubt that such a shift has occurred for Uber within the past 3 years.

As Friedman, Kahn Jr., and Howe point out, events that violate trust may vary in degree. A breach in trust may be repaired, but "betrayals of trust often end relationships" [1 p35]. In the case of our model, events that violate trust will have a greater impact on trust if the company's product or business model relies on a trust relationship. For example, we would expect a larger drop in stock price after a trust violation by Southwest than by a trust violation by another airline. This is a key insight upon which a quantitative measure of trust can be built.

4.2.2   How Trust Violations Impact Companies' Market Value

Given the amorphous and context-dependent nature of trust, it is often easier to measure a change in trust than it is to measure its absolute value. One way of measuring trust in businesses is to examine financial metrics and measures given an event that we know to have caused a change in trust. In our forthcoming paper, "Quantifying the Value of Customer Trust in Companies," we present an analysis conducted by us of 333 crowdsourced examples of trust violations across 15 industry sectors. These violations break down into 11 broad categories of trust violation (e.g., product failure, sexual harassment, data and cybersecurity breaches, labor disputes). We measured the percentage drop in stock price value (e.g., the "signal") after the trust violation, using as the baseline the expected (i.e., "forecast") stock price predicted using data from before the violation (i.e., what the stock price should have been had there been no violation). The blue region indicates the 95% confidence interval for that predicted stock price.

In Fig. 4.1, we see the behavior of the Facebook stock price after it was revealed, in September 2017, that Russian government operatives placed ads on Facebook intended to influence American voters. One can observe Facebook stock prices underperforming with respect to expectation immediately after the scandal broke. However, Facebook shows recovery in the approximately 5 months afterward, suggesting that trust may not be core to its value proposition for users. Stocks drop again in mid-March, when the Cambridge Analytica scandal comes to light. However, again, Facebook stock recovers a few months later.

At some level, Facebook's decisions to share user data with Russian government operatives and with Cambridge Analytica are not surprising. Facebook's business model is centered on selling customer data to advertisers—its main revenue stream comes from its advertisers and not its users. The company is incentivized to commit trust violations because that is how it makes more money. We believe Facebook's stock price shows resilience in the face of trust violations because a lack of trust is already "baked into" its stock price. Due to the nature of its business model, users (and investors) have an expectation that it will violate trust.

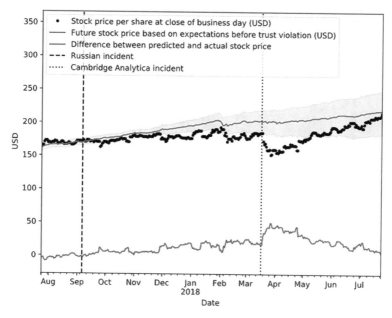

Fig. 4.1 Real vs. Forecast Facebook Stock Price July 2017–July 2018

This same dynamic can be observed in the aggregate. We segmented the cohort of trust violations into those aligned with a company's business model, those not aligned with a company's business model, and those that are ambiguous. We find that the maximum percentage drop in stock price was 14% on average for trust violations aligned with business models, but 22% for those that are not aligned or ambiguous.

Beyond percentage price drops, in "Quantifying the value of consumer trust in companies," we also analyze how long it takes stocks to recover to normal after a trust violation. We find significant differences in average recovery period looking across industry sectors as well as type of violation. The next steps in our line of research include developing natural language processing (NLP) methods to collect data on a much larger cohort of such trust violations (sized in thousands rather than hundreds of trust violations). However, the preliminary data we have already analyzed suggests that some companies derive more of their market value (as measured by stock price) from trust than others [19].

These stock price fluctuations also point to another truth about consumers' trust in companies: Trust is a bandwagon phenomenon. Users' trust attitudes are heavily influenced by those of people around them. This can benefit companies that have a lot of users and/or strong network effects, as it creates inertia that can blunt the impact of individual trust violations. For example, the Cambridge Analytica data breach and other scandals have not caused a mass defection by Facebook users (though there is some evidence that it might have played a role in stagnating user growth) [20]. Even if trust has been damaged, network effects are such that it does not yet show up in numbers. To what level trust needs to be disrupted to overcome network effects (and the benefits of social capital) is a critical open question, both for the specific case of Facebook and also for technologies in general.

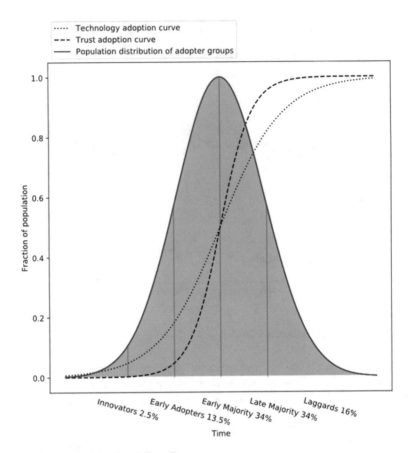

Fig. 4.2 Technology Adoption & Trust S-curves

4.2.3 The Changing Texture of Trust

How does trust in a new company diffuse through such a network and through society in general? The process could follow the pattern defined by the diffusion of innovation theory first put forth by E.M. Rogers in 1962 [21]. Created to explain how new technology gets broadly adopted by a society, the theory places users into five categories depending on the speed with which they adopt a new innovation: innovators, early adopters, early majority, late majority, and laggards. Rogers arranges these groups along a normal distribution so that innovators and early adopters form 16% of the population, the early and late majorities form 34% each, and laggards form another 16%.

Due to these proportions, the rate of adoption of a successful new technology tends to follow what is called an "s-curve" over time. Adoption by innovators and early adopters leads to modest gains in market share, but once a new technology is picked up by an early majority, overall adoption starts to grow rapidly; it then plateaus as it hits saturation with the laggards, the only group left to drive growth (Fig. 4.2).

It seems likely that the deepening of consumer trust in a company follows an s-curve, too—albeit one of a slightly different shape. While there is little empirical data to verify exactly how trust in a company diffuses through society, we can make some theoretical conjectures based on existing research.

In Fig. 4.2, we measure trust in terms of both breadth and depth. The amount of trust measured on the *x*-axis is the product of how many people trust the company as well as the depth of that trust. A company that reaches "maximum" trust along the *y*-axis has built the deepest possible trust relationship it can with the maximum number of people, given the constraints of its business model, history of trust violations, and so on. This "maximum trust" may vary from brand to brand—for example, Facebook's maximum depth of trust is likely less than that of a company like Aspiration, which derives much of its value from trust relationships.

It would be reasonable to assume that increasing trust generally drives adoption by lowering perceived risk. As Florea writes, "High levels of trust [in a brand] automatically mean low perceived risk of adoption [of its innovative product]—as the two concepts are complementary" [22 p1530]. If this relationship were true for all users' trust relationships, and trust in a brand drives adoption in a one directional way, the trust s-curve would simply be the technology-adopted s-curve shifted a bit to the left. Trust in the simplistic case would be a leading indicator of adoption.

However, we would question that hypothesis based on some assumptions about how the different groups of adopters each value trust. Trust in a company may not be very important to an innovator or early adopter whose low levels of risk aversion are easily outweighed by the excitement of trying out new technology. Therefore, we actually expect the rise in trust to lag behind adoption, at least at first (i.e., trust is a lagging signal). However, trust *is* likely more important to members of the early and late majority, who decide to try the new technology based on its existing reputation. Thus, we suggest that at some point at the beginning of these two adoption periods, the trust s-curve crosses the adoption s-curve, and trust begins to drive (and precede) adoption. This transition from lagging to leading mathematically means that the middle of the trust s-curve must be steeper than the equivalent part of the adoption s-curve.

In addition, we hypothesize that this steepness has additional meaning. Trust is based on normative behavior. Once a certain number of users trust a given technology provider, we would expect that trust would diffuse rapidly through the rest of society until it reaches saturation. It is partly a matter of network effects: If someone I trust starts trusting Uber (or Facebook or Lyft) and, therefore, lays the groundwork for building a normative behavior, I am more likely to start trusting that company too. Based on this, we would expect that once the trust s-curve started to climb, it would climb more steeply than an equivalent adoption s-curve.

How, then, does trust in a business actually spread? Behavioral research in marketing and other business disciplines suggests a number of potential vectors. Social science researchers have established the importance of brand community, defined by Muniz and O'Guinn as "a specialized, non-geographically bound community, based on a structured set of social relationships among admirers of a brand" [23 p412]. These communities are defined by a shared consciousness or feeling of belonging. They can be a key vector through which trust in a brand can spread—for example, Apple's strong brand community has been central to promoting the adoption of the iPhone and other innovative products. Social trust in experts can also have a strong influence on adoption, according to Siegrist, Cvetkovich, and Roth, especially when users do not understand the technology well enough to evaluate it themselves. That trust is more likely to grow when those leaders are seen as having similar values [24].

Over the time it takes trust to diffuse, trust might not only broaden, but deepen. Users can be shepherded further down the trust funnel as they become more deeply convinced of a technology provider's ability, integrity, and benevolence. One technology provider may also lay the groundwork for another that pushes the boundaries of user trust even further.

For example, take dog walking: a vulnerable experience that involves letting a stranger into your home when you are not there and trusting them with a pet to which you are emotionally attached. Rover, founded in 2011, is a marketplace app that lets users choose and hire a vetted dog walker based on an online profile [25]. Without Rover's success in building users' trust, it is hard to imagine the popularity of Wag, an app created in 2014 that assigns each user a vetted dog walker on demand via an algorithm, similar to Uber [26]. In just a few short years, users transitioned from needing to choose who would enter their house and walk their dog to allowing the technology provider to choose that person—a clear deepening of trust.

It is also interesting to note that in the past year, these nascent companies have run into troubles that touch on the three core elements of trust. The examples below were highlighted in a September 18, 2018, article on the website Vox [27]:

- **Ability:** Dogs have been lost, injured, or even killed while under the care of Rover and Wag. The Vox article alone lists 14 examples.
- **Integrity:** Dog walkers' training is purportedly minimal and does not involve cardiopulmonary resuscitation (CPR) or first aid. Both companies have been accused of trying to suppress users who complain; Wag sent a cease-and-desist letter to a family that was posting to Facebook complaining the service had lost their dog.
- **Benevolence:** Neither company covers veterinary costs for injured dogs or medical costs for workers injured *by* dogs. One dog walker who asked about coverage during training said she was "laughed at." On the other hand, Wag reportedly spent $30,000 in an attempt to track down one lost dog.

As of this writing, it is unclear what impact, if any, these trust violations will have on Wag and Rover as companies. Unlike the companies described in "Quantifying the value of consumer trust in companies," both Wag and Rover are still private companies and startups. As such, they disclose limited financial data, making it difficult to track the impact of trust violations on business value via the method we identified.

Rover's and Wag's trust relationships with users are in flux partly because IoT as a category is in the midst of a phase transition. Trust in IoT is both broadening—as it extends through the population—and deepening—as users become willing to trust

IoT devices with more sensitive functions. However, the way in which cyberphysical systems bridge the real versus virtual divide introduces some interesting differences in how trust might influence adoption in a way that is different than for purely digital technologies.

4.3 Applying Trust in Companies to IoT

In the previous section, we slipped easily from talking about the stock price of Facebook to the evolution of trust in Rover and Wag. However, there are important distinctions between a purely digital product like Facebook and the kind of cyberphysical systems we will discuss here.

We have mentioned that Facebook may not suffer major loss of market value in the face of trust violations in part because a lack of trust is baked into its stock price. There may also be another reason, which is that the impacts of trust violations in the purely digital realm remain relatively abstract. When Uber violates trust, users may fear for their physical safety. By contrast, when Cambridge Analytica steals and abuses Facebook user data, the only direct impact is that users see different targeted ads.

Phelan, Lampe, and Resnick posit that individuals' response to potential privacy incursions online actually have two elements that may be in conflict with each other: the intuitive or emotional response and the deliberate or considered response. These two types of response correlate roughly with the intuitive, unconscious "System 1" and intentional, conscious "System 2" of the dual process model of cognition, which was popularized by Daniel Kahneman in his 2011 book, "Thinking, Fast and Slow" [28]. When an individual does not intuitively find a privacy incursion "creepy," they may bypass the more deliberate process of constructing a rational objection and taking action [29].

This is one possible explanation for the so-called "privacy paradox"—the fact that people claim they want more privacy online but then do not take action to secure that privacy [30]. Human beings' privacy instincts evolved in the physical, not the digital, world. We respond intuitively to physical intrusion—a face peering in our window or a stranger in our house. Sources as ancient as the Talmud recognize a right to privacy in the home [31]. However, our human instincts and cultural precepts may be less fully developed when it comes to digital incursions [32]. A desire for "privacy" in cyberspace is in most cases simply a desire for control—specifically, control over how our online data is used. Yet most of us continue to participate in online spaces where we are denied that control.

In contrast to those purely digital spaces, cyberphysical systems bring the impact of trust violations into "the real world." An improperly vetted Wag dog walker could end up losing or injuring your beloved pet. A malfunction with the Airbnb app could lock you out of your lodging and make for a very physically uncomfortable night. We have already mentioned a few examples of Uber's trust violations that have included physical assault on a user.

All of this should make adoption of IoT and IoT-based services feel inherently risky to users. It should, in turn, make trust in technology providers more important, with trust violations generating strong and instinctive negative reactions from users. As such, you would expect trust to be an important part of an IoT company's brand value, with large, lasting fluctuations in stock price when that trust was violated.

However, research shows that this is mostly not the case, at least not yet. In the early stages of IoT proliferation, consumers actually value **ability**, the lowest level of trust, most highly with IoT products. Bai and Gao were surprised to find, in their survey of 368 Chinese consumers' likelihood to adopt electronic toll collection (ETC) technology, that trust actually had relatively little impact. They suggested that familiarity could be shaping the trust relationship:

> One explanation for the insignificant effect of trust on behavioral intention lies in the lack of the interaction between consumers and IoT devices/systems. IoT technologies are relatively new in China, and as a result, users' familiarity with these technologies and their relevant products is low. For instance, many people may not have the basic knowledge of how to use the ETC and how it operates. Consumers hardly know this technology; thus, they are not willing to assess whether this type of technology is secure or trustworthy.

This suggests that—as IoT devices become more widespread and accepted, increasing users' familiarity with basic knowledge about the technology—trust may begin to play a role in their adoption by creating normative behavior in the way that we would expect [33].

Ultimately IoT makes consumers more *vulnerable* than purely digital systems like Facebook. IoT companies that are able to leverage those vulnerabilities to shift to higher levels of trust will differentiate themselves and build more durable relationships with customers.

Also, while consumers may become more familiar with a particular IoT technology, they may still not actually understand *how it works* in the sense that they understand how a hammer works. In that case, the trust they maintain in a particular technology once they are aware of its potential impacts will be strongly rooted in their trust in the technology provider. Future researchers into trust in IoT technologies should keep this link in mind and use the ability, integrity, benevolence framework to evaluate trust in the technology provider as well as the technology itself. We do so in the following section for three important IoT technologies emerging in the market today.

4.3.1   Three Case Studies on Trust in IoT

As discussed previously, human beings may have very different reactions to violations that impact the physical compared to the digital realm. Thus we have divided potential questions about each product or service's ability, integrity, and benevolence along these lines as well. In our discussion of each example, we attempt to tease out which of these questions address the technology itself, the company behind it, or both. We hope that these examples will illustrate for future researchers how our framework might be applied.

4.3.1.1   Bird Scooters

Founded in 2017, Bird (an Alpha Edison portfolio company) is a scooter-sharing service that operates in 100+ cities including Los Angeles, California; Nashville, Tennessee; Washington, DC; and Atlanta, Georgia [34]. Unlike the bike-sharing services common in many major cities, Bird allows users to pick up and drop off their vehicles anywhere, without the need to find a set docking station. Contract workers pick up the scooters at night for recharging [35]. In June 2018, Bird became the fastest company ever to achieve a $1 billion valuation, earning it "unicorn" status alongside companies like Airbnb and Uber [36]. But do users trust this new company that has suddenly dropped scooters on the streets of their town? And how much does that trust matter (Table 4.2)?

In general, it seems likely that trust in the company will be a less important driver of adoption for Bird than it is for many other IoT products. With ride-hailing services, for example, users must trust a stranger hired by the ride-hailing company with their life and limb. By contrast, since Bird users drive the scooters themselves, their physical safety is largely their own responsibility (as long as the scooter is in good repair).

However, trust can come into play in a more abstract way. For example, here the ability and integrity questions address an intersection of technology-related and brand-related concerns. The question of whether Bird keeps users' data private, for example, is both a question about the app (is it built with adequate protections against hackers looking to steal credit card numbers?) and a question about the company (does Bird value users' privacy enough to develop and enforce strong data management policies?). Trust in the company may also come into play in the physical realm. If a user sees Bird scooters littering their sidewalks as a major inconvenience, they may impute a lack of integrity to Bird the company for allowing this to happen. Most of these questions, however, have lower stakes than the others we will treat in this section.

4.3.1.2   Amazon Alexa

Alexa is a virtual personal assistant designed and built by Amazon. Like other virtual assistants, she is capable of understanding and responding to spoken human language. While Alexa is most strongly associated with the Amazon Echo smart

Table 4.2 Questions About Trust in Bird

	Physical	Digital
Ability	• Will scooters be available where I am when I need them?	• Is it easy to reserve a scooter on the app?
Integrity	• Are the scooters safe? • Are the scooters left on sidewalks, etc. an inconvenience?	• Is my ride data private? • Will the company refund my ride if I experience a problem?
Benevolence	• How will I be treated if I have an accident?	• Are Bird workers well treated?

Table 4.3 Questions About Trust in Amazon

	Physical	Digital
Ability	• Will Alexa do what I ask her to?	• Is Alexa "smart" enough to answer my questions?
Integrity	• Are products sustainable or ethically made?	• Are my conversations with Alexa private?
Benevolence	• Does Alexa increase quality of life for users?	• Will the information gathered via Alexa be deployed in the user's best interests?

speaker, Amazon currently lists dozens of devices that can connect to Alexa, including light bulbs and microwaves [37]. While individual sales figures on Alexa devices are not available, according to the technology research firm Canalsys, the number of smart speakers installed worldwide reached 40 million by the end of 2017 [38]. Users trust the product enough to put a live microphone in their personal spaces, but do they feel they have sufficient control of the data or receive benefits that are worth the trade-off (Table 4.3)?

Unlike in the previous example, the benevolence questions here are less tightly tethered to Amazon's brand. The question of whether Alexa increases quality of life, for example, depends more on the technology's impact: Does having a virtual assistant free up the user's time for more fulfilling tasks? Does it reduce loneliness by giving them someone to talk to? Does it connect them to the internet without isolating them from loved ones in the room (as smartphones have been thought to do)?

In part because Alexa operates in such a protected, private space—the home—we expect trust in Amazon as a company to be an important factor in Alexa adoption. As with Bird, the integrity question of how private a user's data remains is a blend of technology- and brand-related concerns. However, the benevolence question of how Amazon will use that data is manifestly a question about the company. Suspicious users have suspected Alexa of "spying on them" from day one, and there have been a few documented cases of Alexa recording snippets of conversation when she had not been awakened by her "wake word" [39]. What you believe Amazon does with such recordings—accidental or not—likely depends on how much you trust it as a company. In turn, your trust in the company is likely to influence your adoption of Alexa.

4.3.1.3 Tesla Self-Driving Cars

Depending on which experts you ask, self-driving cars may be on the road in 1 year or not for decades [40]. However, they yield an interesting hypothetical example for this chapter because of how much their success will rest on the development of trust. Behavioral research on "algorithm aversion" has suggested that humans tend to trust recommendations from automated systems less readily than they do human judgement [41]. While widespread adoption of self-driving cars is expected to significantly reduce traffic accidents, most users are still skeptical of the technology and more trusting of human drivers. Widespread adoption of self-driving cars will require a massive trust leap. Some commentators have suggested Uber could suffer a disadvantage in the future market for self-driving cars for exactly this reason: It is a market where only a very trusted company could succeed [42].

While many companies are developing autonomous vehicles, here we have chosen to focus on the electric car company Tesla. An April 2018 survey by car shopping website Autolist found that 32% of consumers trusted Tesla to bring a self-driving vehicle to market, more than double the percentage of those who trusted the runner-up, Toyota [43]. Tesla's existing vehicles are already equipped with Autopilot, a feature that turns on some self-driving capabilities, but is only supposed to be used when the driver has their eyes on the road and their hands on the wheel.

However, as of this writing, at least two people have died while using Autopilot [44]. The erratic behavior of CEO Elon Musk and the departure of multiple top executives from Tesla has also shaken trust in the company more broadly [45]. Trust in Tesla is a life or death decision for consumers and, therefore, the most complex of our three examples (Table 4.4).

It is easy to see ways in which trust in Tesla could play into the questions listed above. Thanks to intensive media coverage, the general public is much more familiar with how self-driving cars are supposed to "think" than they are for other sorts of artificial intelligence (AI), so they may be likely to ask questions about ethical decision-making on the road. Since it is impossible for every driver to evaluate the algorithm used by each car on the market (both due to lack of expertise and the fact that such algorithms tend to be uninterpretable black boxes even to experts), trust in a technology provider will be a particularly important way to build this trust bridge. For early adopters of self-driving cars, Musk's reputation as an inventor of and investor in ambitious new technologies (he is also founder of space transportation company SpaceX) may be a mark in Tesla's favor.

Nonetheless, Tesla will have to overcome the difficulties mentioned above as well as the perception that it is not a company that cares deeply about safety. In 2018, one of Tesla's plants was investigated by California's job safety agency due to

Table 4.4 Questions About Trust in Tesla

	Physical	Digital
Ability	• Can the car get me from point A to point B? • Does it do so consistently without crashing?	• Does the car navigate properly?
Integrity	• If I am injured in an accident, how will I be treated?	• What protections are in place to keep the car from getting hacked? • How does the car make ethical decisions—e.g., whether to hit a squirrel or swerve and hit a human?
Benevolence	• Do self-driving cars reduce accidents and make roads safer?	• Will Tesla give police control of my self-driving car when an officer wants to pull me over? • Will Tesla share my private data with law enforcement upon request?

a frequency of worker injuries that "exceeded the industry average" as well as allegations that some injuries were not reported at all, as is legally required. It remains to be seen whether these and other issues impact Tesla's ability to lead the burgeoning self-driving car industry. Of all the examples discussed in this section, this is the one with the highest stakes: Users' lives literally depend on whether Tesla is trustworthy.

4.4   Applications to Venture Capital

Venture capitalists must be prognosticators. In order to generate strong returns on our investments, we seek out solutions that are not yet obvious to the rest of the market and then invest in startups working in those spaces before others realize the opportunity is there.

Our interest in the area of trust derives from our practical need to identify the most promising opportunities for investment, which include the many new companies that are trying to reinvent trust relationships with customers. Companies that deploy trust as a core asset do not just stand to build stronger customer relationships and edge incumbents out of existing markets—they can build entirely new markets by tapping into latent demand [46].

We see this trend as particularly important in what we would consider the IoT space, which is dominated today by small, new companies. In the successes of Uber, Lyft, Airbnb, and other companies in the so-called "sharing economy"—many of which blend online elements with real-world services—we can already see the beginnings of a trust revolution. By applying the framework we have laid out in this chapter and other publications, we can become more deliberate about finding these and other investment opportunities early on.

Still, as mentioned previously, quantifying the value of trust for privately held startups is challenging. With a public company, it is possible to follow fluctuations in market value day-to-day or even minute-by-minute on a public exchange, making it relatively easy to establish the impact of a given trust violation. For a private company—especially a very small, new one—there is comparatively little data. How do we understand the trust relationship a company is building with users before they have many users, much less a stock market price?

This is why, in the IoT-specific section of this chapter, we have focused on developing our qualitative framework for understanding the changing texture of trust, rather than analyzing quantitative data. In the absence of hard numbers on a particular startup, a conceptual understanding of how trust relationships could or should develop is often our main guide for establishing the scope of the opportunity for us as investors.

As the company grows and more data become available, we expect the trends we forecasted initially to become more obvious over time, corroborating our hypothesis. Similarly, we hope that with time, more data will become available to verify the broader thesis we have advanced about trust in IoT companies here. In the meantime, we hope that our framework will be useful to researchers across disciplines interested in looking at trust in IoT technology in a more holistic way.

Acknowledgments This research was supported by Alpha Edison. We are thankful to Nathalie Lagerfeld for contributing significantly to the writing of this chapter and Kim Ledgerwood for improving the form of the manuscript. We thank Sam Bogen for his assistance on the trust violation analysis. We also have to express our appreciation to Nate Redmond for sharing his pearls of wisdom with us during the course of this research.

References

1. Friedman B, Kahn PH Jr, Howe DC (2000) Trust online. Commun ACM [Internet] 43(12):34–40 [cited 16 Nov 2018]. Available from: https://www.researchgate.net/profile/Daniel_Howe3/publication/220420837_Trust_Online/links/571f02f508aefa648899aa13.pdf.   https://doi.org/10.1145/355112.355120
2. Botsman R (2017) Who can you trust?: how technology brought us together–and why it could drive us apart. Penguin, UK
3. Molla R (2018) Americans seem to like ride-sharing services like Uber and Lyft. But it's hard to say exactly how many use them. Recode [Internet]. Vox Media, Washington, DC [cited 16 Nov 2018]; [about 4 screens]. Available from: https://www.recode.net/2018/6/24/17493338/ride-sharing-services-uber-lyft-how-many-people-use
4. Barinka A, Newcomer E (2018) Uber valued at $120 billion in an IPO? Maybe. Bloomberg [Internet] [cited 16 Nov 2018]; Technology: [about 4 screens]. Available from: https://www.bloomberg.com/news/articles/2018-10-16/uber-valued-at-120-billion-in-an-ipo-maybe
5. Crunchbase [Internet]. San Francisco: Crunchbase Inc. c2018 [cited 16 Nov 2018]. Uber; [about 15 screens]. Available from: https://www.crunchbase.com/organization/uber
6. Susan LS (2018) Fowler's plan after Uber? Tear down the system that protects harassers. Guardian [Internet] [cited 16 Nov 2018]; Technology:[about 5 screens]. Available from: https://www.theguardian.com/technology/2018/apr/11/susan-fowler-uber-interview-forced-arbitration-law
7. O'Brien SA, Black N, Devine C, Griffin D (2018) CNN investigation: 103 Uber drivers accused of sexual assault or abuse. CNN [Internet]. Cable News Network, New York [cited 16 Nov 2018]; [about 11 screens]. Available from: https://money.cnn.com/2018/04/30/technology/uber-driver-sexual-assault/index.html
8. Friedman B (ed) (1997) Human values and the design of computer technology. Cambridge University Press, New York
9. Friedman B, Kahn PH Jr, Borning A (2002) Value sensitive design: theory and methods. Seattle, WA: University of Washington: UW CSE Technical Report 02-12-01. Supported by NSF Awards IIS-9911185, SES-0096131, EIA-0121326, and EIA-0090832. Available from: https://faculty.washington.edu/pkahn/articles/vsd-theory-methods-tr.pdf
10. Manders-Huits N (2011) What values in design? The challenge of incorporating moral values into design. Sci Eng Ethics 17(2):271–287. https://doi.org/10.1007/s11948-010-9198-2
11. McKnight DH, Cummings LL, Chervany NL (1998) Initial trust formation in new organizational relationships. Acad Manag Rev 23(3):473–490.. Available from: https://pdfs.semanticscholar.org/8439/27d55bf91abc21961a1f9435e31ec3e92326.pdf.   https://doi.org/10.5465/amr.1998.926622
12. Schlosser AE, White TB, Lloyd SM (2006) Converting web site visitors into buyers: how web site investment increases consumer trusting beliefs and online purchase intentions. J Mark 70(2):133–148. https://doi.org/10.1509/jmkg.70.2.133

13. Siegrist M (2000) The influence of trust and perceptions of risks and benefits on the acceptance of gene technology. Risk Anal 20(2):195–203. https://doi.org/10.1111/0272-4332.202020
14. Hinchliffe E (2017) Check the numbers, it's true: People really like Lyft over Uber, even before #DeleteUber. Mashable [Internet]. Mashable, New York. [cited 16 Nov 2018]; [about 4 screens]. Available from: https://mashable.com/2017/03/21/lyft-ipsos-data/#V84nKXlPrPq3
15. McKnight DH, Choudhury V, Kacmar C (2002) The impact of initial consumer trust on intentions to transact with a web site: a trust building model. J Strateg Inf Syst 11(3–4):297–323. Available from: https://msu.edu/~mcknig26/TrBldgModel.pdf
16. Prusak L, Cohen D (2001) How to invest in social capital. Har Bus Rev [Internet] [cited 16 Nov 2018]; [about 18 screens]. Available from: https://hbr.org/2001/06/how-to-invest-in-social-capital
17. Aspiration [Internet]. Marina Del Rey: Aspiration Partners c2018 [cited 16 Nov 2018]. Who we are; [about 4 screens]. Available from: https://www.aspiration.com/who-we-are/
18. Redmond N, Danneman B. Trust: building intimacy at scale. Forthcoming
19. Zhao F, Danneman B. Quantifying the value of consumer trust in companies. Forthcoming
20. Constine J 2018 Facebook shares climb despite Q3 user growth and revenue. TechCrunch [Internet]. Oath Tech Network, San Francisco; [cited 16 Nov 2018]; [about 6 screens]. Available from: https://techcrunch.com/2018/10/30/facebook-earnings-q3-2018/
21. Rogers EM (1995) Diffusion of innovations, 4th edn. The Free Press, New York
22. Florea DL (2015) The relationship between branding and diffusion of innovation: a systematic review. Procedia Econ Finance 23:1527–1534. https://doi.org/10.1016/S2212-5671(15)00407-4
23. Muniz AM Jr, O'Guinn TC (2001) Brand community. J Consum Res [Internet] 27(4):412–432. [cited 16 Nov 2018]. Available from: https://www.jstor.org/stable/10.1086/319618?seq=1#metadata_info_tab_contents
24. Siegrist M, Cvetkovich G, Roth C (2000) Salient value similarity, social trust, and risk/benefit perception. Risk Anal 20(3):353–362. https://doi.org/10.1111/0272-4332.203034
25. Crunchbase [Internet]. San Francisco: Crunchbase Inc. c2018 [cited 16 Nov 2018]. Rover; [about 15 screens]. Available from: https://www.crunchbase.com/organization/rover-com
26. Crunchbase [Internet]. San Francisco: Crunchbase Inc. c2018 [cited 16 Nov 2018]. Wag; [about 14 screens]. Available from: https://www.crunchbase.com/organization/wag
27. Lieber C (2018) The startup world's cuddly, cutthroat battle to walk your dog. Vox [Internet]. Vox Media, Washington, DC; [cited 16 Nov 2018]; [about 22 screens]. Available from: https://www.vox.com/the-goods/2018/9/12/17831948/rover-wag-dog-walking-app
28. Kahneman D, Egan P (2011) Thinking, fast and slow. Farrar, Straus and Giroux, New York
29. Phelan C, Lampe C, Resnick P (2016) It's creepy, but it doesn't bother me. In: Mental models of privacy. CHI conference on human factors in computing systems; 2016 May 7–12; San Jose, CA. New York: ACM; 2016. Available from: http://www-personal.umich.edu/~cdphelan/files/p5240-phelan.pdf. https://doi.org/10.1145/2858036.2858381
30. Barth S, de Jong MD (2017) The privacy paradox—investigating discrepancies between expressed privacy concerns and actual online behavior—a systematic literature review. Telematics Inform [Internet] 34(7):1038–1058 [cited 16 Nov 2018]. Available from: https://www.sciencedirect.com/science/article/pii/S0736585317302022
31. Enkin A (2012) Privacy [Internet]. [cited 16 Nov 2018]. Legacy: Torah Musings: [place unknown: Torah Musings. [about 2 screens]. Available from: https://www.torahmusings.com/2012/07/privacy/
32. Zhao F (2018) Will smart home tech make us care more about privacy? TechCrunch [Internet]. Oath Tech Network, San Francisco; [cited 2018 Nov 16]; [about 6 screens]. https://techcrunch.com/2018/06/03/will-smart-home-tech-make-us-care-more-about-privacy/
33. Gao L, Bai X (2014) A unified perspective on the factors influencing consumer acceptance of internet of things technology. Asia Pac J Mark Logist 26(2):211–231
34. Bird [Internet]. [place unknown]: Bird. [cited 16 Nov 2018]. Cities; [about 7 screens]. Available from: https://www.bird.co/cities/

35. Lorenz T (2018) Electric scooter charger culture is out of control. Atlantic [Internet]. The Atlantic Monthly Group, Washington, DC; [cited 16 Nov 2018]; [about 12 pages]. Available from: https://www.theatlantic.com/technology/archive/2018/05/charging-electric-scooters-is-a-cutthroat-business/560747/

36. Griswold A (2018) Bird is the fastest startup ever to reach a $1 billion valuation. Quartz [Internet]. [place unknown]: Quartz; [cited 16 Nov 2018]; [about 4 screens]. Available from: https://qz.com/1305719/electric-scooter-company-bird-is-the-fastest-startup-ever-to-become-a-unicorn/

37. All things Alexa [Internet]. [place unknown]: Amazon.com; c1996–2018 [cited 16 Nov 2018]. Amazon Echo and Alexa Devices; [about 6 screens]. Available from: https://www.amazon.com/Amazon-Echo-And-Alexa-Devices/b?ie=UTF8&node=9818047011

38. Shulevitz J (2018) Alexa, should we trust you? Atlantic [Internet]. The Atlantic Monthly Group, Washington, DC; [cited 16 Nov 2018]; [about 36 screens]. Available from: https://www.theatlantic.com/magazine/archive/2018/11/alexa-how-will-you-change-us/570844/

39. Murdock J (2018) Is Alexa spying on you? Amazon responds after rogue Echo device leaks couple's private chat. Newsweek [Internet]. Newsweek, New York; [cited 16 Nov 2018]; [about 3 screens]. Available from: https://www.newsweek.com/alexa-spying-you-amazon-responds-after-rogue-device-secretly-records-private-944557

40. Kessler S (2017) A timeline of when self-driving cars will be on the road, according to the people making them. Quartz [Internet]. [place unknown]: Quartz; [cited 16 Nov 2018]; [about 10 screens]. Available from: https://qz.com/943899/a-timeline-of-when-self-driving-cars-will-be-on-the-road-according-to-the-people-making-them/

41. Dietvorst BJ, Simmons JP, Massey C (2015) Algorithm aversion: people erroneously avoid algorithms after seeing them err. J Exp Psychol Gen [Internet] 144(1):13 p. [cited 16 Nov 2018]. Available from: http://opim.wharton.upenn.edu/risk/library/WPAF201410-AlgorthimAversion-Dietvorst-Simmons-Massey.pdf

42. Siddiqui F (2017) #DeleteUber will have lasting fallout for ride-hailing app, study says. The Washington Post [Internet]. [cited 16 Nov 2018]; Gridlock: [about 6 screens]. Available from: https://www.washingtonpost.com/news/dr-gridlock/wp/2017/05/16/deleteuber-will-have-lasting-fallout-for-ride-hailing-app-study-says/?utm_term=.cf5de7ed6b96

43. Elmerraji J (2018) In Tesla we trust, new study reveals. The Street [Internet]. The Street, Inc., New York; [cited 16 Nov 2018]; [about 5 screens]. Available from: https://www.thestreet.com/investing/in-tesla-we-trust-new-study-reveals-14566222

44. Stewart J (2018) Tesla's self-driving autopilot was involved in another deadly crash. Wired [Internet]. Condé Nast, New York; [cited 16 Nov 2018]; [about 7 screens]. Available from: https://www.wired.com/story/tesla-autopilot-self-driving-crash-california/

45. Hull D (2018) 'There's something wrong': Tesla's rapid executive turnover raises eyebrows as Musk thins the ranks. Financial Post [Internet]. Post Media Network, Toronto; [cited 16 Nov 2018]; [about 6 screens]. Available from: https://business.financialpost.com/transportation/autos/theres-something-wrong-rapid-tesla-executive-turnover-raises-eyebrows-as-musk-thins-the-ranks

46. Redmond N (2018) Why most investors get market size wrong over and over again. Forbes [Internet]. Forbes Media, LLC, New York; [cited 16 Nov 2018]; [about 8 screens]. Available from: https://www.forbes.com/sites/valleyvoices/2018/05/04/why-investors-get-market-size-wrong/#56fe54105b83

Chapter 5
Ahead of the Curve: IoT Security, Privacy, and Policy in Higher Ed

Joanna Lyn Grama and Kim Milford

5.1 IoT on Campus

Picture a college or university. Images that spring to mind quickly are a bustling urban campus with modern buildings or a bucolic campus featuring an abundance of limestone and red brick. One might also think of technology-enabled classrooms, cutting-edge research labs, and innovative makerspaces. Technology has pervaded today's institutions of higher education. And it is not just that colleges and universities themselves are using technology that is owned by the institution. Students, faculty, staff, and visitors are bringing more personal devices than ever before to campus. In fact, the rise of *"Bring Your Own Everything"* (BYOE) technology in higher education has been well documented, as early as 2013 [1]. Consider for example, today's student population [2]:

- 66% of today's undergraduate students own two or three Internet-capable devices and 78% of students connect those devices to the campus network simultaneously.
- 52% of students rate campus network performance (e.g., high speed and no interruptions) as good or excellent.

The Internet of Things (IoT) represents the evolution of the BYOE construct in higher education. While BYOE acknowledged the increasing number of Internet-capable devices brought on campus, IoT refers both to the increasing number of

J. L. Grama (✉)
Vantage Technology Consulting Group, El Segundo, CA, USA
e-mail: Joanna.grama@vantagetcg.com

K. Milford
Research and Education Networks Information Sharing and Analysis Center (REN-ISAC),
Bloomington, IN, USA
e-mail: kmilford@ren-isac.net

© Springer Nature Switzerland AG 2019
F. D. Hudson (ed.), *Women Securing the Future with TIPPSS for IoT*, Women in Engineering and Science, https://doi.org/10.1007/978-3-030-15705-0_5

devices and the understanding that those devices are "connected." Fall semester 2018 saw the first institutional provisioning of IoT devices to all students on campus, with the rollout of the 2300 Echo Dots at Saint Louis University [3]. This heralds a new era of IoT deployment at universities with new issues beyond BYOE. The IoT is characterized by widely available consumer Internet connectivity, decreasing consumer cost for that connectivity, lower technology costs in general, the increasing number of devices that are Internet-enabled out of the box, and the large amounts of data generated by those devices that is stored "in the cloud." And, the IoT is growing. Internet-connected things, from PCs to tea kettles, number in the billions [4], and every day 44 exabytes of data are created [5].

The higher education culture, which is focused on learning and innovation, with a profound emphasis on openness and transparency, has allowed the proliferation of Internet-capable devices to flourish and thrive on today's campuses. Almost every student (98%) and faculty member (96%) have a smartphone and/or tablet (98%), so managing this level of mobility is of key import to higher education [6]. As is the case with IoT, managing mobility does not just refer to devices themselves. It also means managing the mobility of people and their data, and the institution and its data.

Higher education's early experience with BYOE provides it with the building blocks needed to more fully address the promise and peril of the Internet of Things (IoT) in the campus environment. This chapter presents some of the security, privacy, and infrastructure issues that the proliferation of mobile and connected devices bring to campus and how US higher education institutions are responding to the complexities—opportunities and challenges—presented via the rise of the Internet of Things.

5.2 IoT Threats to Privacy and Cybersecurity

The state of the hack has changed considerably in recent years. While the drumbeat of high-profile data exposures and hacks continues with little abatement, see Fig. 5.1, the techniques, tactics and practices (TTPs) of malicious actors grow and evolve. One credential or device is now manipulated to yield access to multiple applications, including stores of personally identifiable information. A single public IP address is used to expose the whole network, providing valuable architecture information that can then be used to bore deeper into organizational resources. Phishing, the most common initial attack vector, has evolved to look much more credible, tailoring messages to target specific victims. The largest denial of service attack to date was reported in March 2018, in which GitHub was targeted with 1.3 Terabits Per Second (Tbps) of sustained Internet traffic for 8 min [7]. Most online attacks to date have been committed by cyber-criminals, exploiting human and system vulnerabilities for financial gain.

Higher Education sees similar TTPs and relative numbers of attacks as other sectors, essentially providing a microcosmic view of cyberthreats on the Internet. As

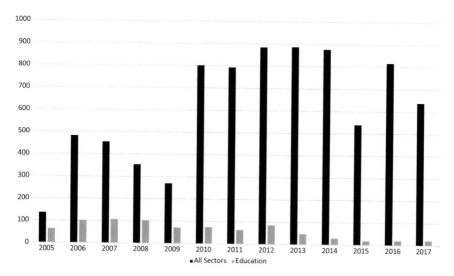

Fig. 5.1 Number of Data Breaches in the USA (Source: Privacy Rights Clearinghouse)

reported by the Privacy Rights Clearinghouse and evidenced in Fig. 5.1, the raw number of US data breaches in the education sector are small in comparison to other industries, although the same types of breaches are experienced, including phishing, malware, and stolen credentials [8].

There are many ways in which the Internet of Things increases risks. It is often the very convenience of the devices that makes them more vulnerable. Lightweight authentication and authorization practices make smart devices easier to access—for legitimate users as well as attackers and criminals. Unencrypted transmission to and from networked devices exposes data transmissions to spying and capture. Device software is built on operating systems that become defunct and unsupported as they age, unbeknownst to the device owners or users. Default system configurations that ease the connection burden for owners generally entail weakened controls. Who does not love the ability to connect without password resets? Many network-aware devices can also store and transmit personally identifiable data about the owner back to the manufacturer without the owner's knowledge or consent.

The range of network-aware devices continues to grow. It has been posited that there were 23 billion IoT connected devices installed as of the end of 2018, and the number could rise up to 75 billion or higher by 2025 [9]. Medical IoT devices show great promise and market growth. Personal health monitoring devices, such as glucometers, scales, heart rate and blood pressure monitors allow patients and their health care professionals to track vital health signs in real time. As a part of its "Whole Person Education," Oral Roberts University implemented mandatory Fitbit use for freshmen [10]. Wearable devices allow for immediate urgent assistance at the push of a button. Talking medical devices remind patients to take medication, check blood pressure, or get exercise. Pacemakers and insulin pumps provide feedback to health providers about usage and patient needs.

Chatty medical devices exacerbate privacy threats inherent in all IoT devices. Personal health information is highly regulated private data, and electronic health records are valuable on the black market [11]. Yet these innovative devices track increasingly greater amounts and types of health information, readily available and transmitted automatically to electronic medical record stores. It is not much of a stretch to imagine that these stores of personal health data might be linked together to provide a comprehensive picture of individuals' health in the near future. Not only will that create risk to protected health information, the compilation of the data may increase risks to all personally identifiable data, as conditions, symptoms, and personal characteristics combine to create digital markers about an individual.

The Mirai attack in 2016 provides a case study in IoT threats, in which the weaknesses inherent in default password use and unpatched operating systems of home routers and IP cameras were exposed. The perpetrators used these common IoT vulnerabilities to exploit devices with the Mirai malware, conscripting the devices to generate massive network activity aimed at popular web sites, and even renting their botnet army out to others. Even with this high-visibility attack, there has been little increase in deploying security protections for protecting IoT devices, and organizations have done little to reduce the likelihood or impact of future attacks driven by IoT devices. In March 2018, Research and Education Networks Information Sharing and Analysis Center (REN-ISAC) notified higher educational institutions in the United States about 16,000 suspected compromised machines. The Mirai malware was the fifth most common exploit seen.

Another major impact in IoT is the ability to perform data analytics on the data flooding in from connected devices. Data can be aggregated from multiple sources to provide millions of data points about consumers. Using increasingly sophisticated data analytics software, data can be leveraged to draw conclusions about individuals and behaviors, even going so far as to predict future events. Analysis derived from IoT-device data can help businesses to pinpoint consumer needs, providing improved products and services. Consumers can get smarter about their own activities, with data analytics providing friendly dashboards, often comparing them to others. Video analytics can be used to improve safety.

All of these innovations in analytics come at a cost: decreased individual privacy. The data used to develop business and consumer dashboards is rich in its detail, which could include surveillance. The time someone wakes up in the morning can be tracked by their smart home speaker, refrigerator, and fitness band. Their morning commute is reported by their car. A trip to the coffee shop is recorded by a camera. A stop at the grocery store on the way home from work, at which the individual uses their store savings card, yields more information about that individual— what time they shopped and what they bought. Many consumers realize this tradeoff, but most have no idea of the volume of information they trade for convenience. Nor are they aware of the data analytics working behind the scenes to tie all the data together into patterns, trends, and predictions.

IoT devices could also have significant impact on learning analytics in the higher education environment. Devices can track and report student behavior valuable in designing learning tools and increasing the effectiveness of teaching practices. As institutions store and analyze more of this data for beneficial purposes, policy consideration for the privacy of the data and the usage of the analytics are needed. While 74% of faculty expressed confidence in their institution's ability to safeguard student information and 65% of faculty have confidence in their institution's ability to safeguard research data, current safeguards and processes may not be enough [12].

5.3 IoT Impacts on Campus Infrastructure

Students, faculty and staff who bring IoT devices to campus or use those provided raise resource management issues. Additional connected devices increase network traffic. Without advanced capacity planning, additional traffic can slow access to mission-critical teaching, learning, and clinical systems. Additional traffic from unknown sources—activities emanating from consumer devices—can obfuscate malicious traffic and introduce abnormalities in network monitoring. New types of traffic unfamiliar to network operating centers (NOCs) and security operating centers (SOCs), can make it difficult to detect aberrant traffic and prevent malicious activity.

No current research book would be complete without addressing the impact of cloud services on information technology. Managing complex storage, transmission, and usage needs with cloud providers requires a full-time-equivalent professional position at many higher educational institutions. The risks of cloud provisioning are not necessarily higher or lower than on-premise IT, but they are different. Cloud providers represent a new breach vector, a technical environment outside the physical boundaries of the campus. Cloud management responsibilities include legal analysis, contract negotiation, technology architectural planning, security assessment, policy drafting, and communications.

All that data created by IoT devices is stored somewhere, and that somewhere is generally the cloud. This makes it easier to provide the consumer with feedback about their device and its usage—anytime, anywhere. In exchange for using the device and its associated cloud storage, consumers may agree to allow their data to be accessed, harvested, and analyzed. With click-through agreements governing the relationship, consumers may never know the physical location of the data storage, what protections exist for personally identifiable data, and what happens to their data over time.

In the classic BYOE environment, the institution is not a party to the agreement between the consumer and the provider. The institution likely does not even know about it. And yet, the institution has a role in transmitting the data to/from the consumer via the campus network. It is as yet unknown where responsibility begins and ends legally and practically for data derived, created, and transmitted by IoT devices. A student consumer acquiesces to an agreement with an IoT device provider, then uses resources at their college to transmit and access data created by the device in the cloud. If the data is breached at the storage site that seems to be under the responsibility of the IoT device/cloud service provider. If the data is breached in transit, the answer may not be so clear. Responsibility could be placed on the provider, or on the college or university acting as a conduit.

In their fall 2018 institutional deployment, Saint Louis University smartly side-stepped the privacy issue by enabling only general applications (or "skills") which do not require the Echo devices to know individual students [3]. That comes at a cost, though, as it curtails many helpful functions on the devices. Students may find they enjoy the convenience of the personally customized device and demand that from their institutions. When or if this happens, it will not change the types of risk, but it could greatly impact the scale and the burden of responsibility assumed by the institution.

5.4   IoT Impacts on Campus Policy

Institutions use their IT policies, and supporting standards, guidelines, and proce-dures, to set forth a clear plan for how they manage IT resources and the data con-tained in those resources in order to support the various missions of the institution. They also use policies to show compliance with laws and regulations affecting IT resources and data. Many frameworks are available to help institutions create their IT policies, but in general the higher education policy development process is one marked by continuous maintenance and review. Unlike other industry sectors, insti-tutional IT policies exist within an environment that values learning, outreach, and innovation. This environment requires certain flexibility in the policy-making process.

From a policy-making perspective, one-size-fits-all simply does not apply in the campus environment. Institutions are not homogenous in form and function, and in fact many campuses function like small cities. Some have their own police depart-ments, health centers and hospitals, power plants, airports, and conference and event services, leading to a very complex and highly integrated IT environment. The complexity does not end there. Some institutions are commuter campuses with no residential students, while others have large populations of student residents. That means that the campus IT infrastructure is used to provide both institutional busi-ness functions (e.g., teaching, research, and administrative service) but also residen-tial (e.g., consumer) functions in dormitories, residence halls, or temporary faculty and staff housing. The complexity in the campus policy environment is further exac-erbated by IoT growth on campus.

5.4.1   IoT in the Office

Two different types of policies have emerged as critical for managing the technol-ogy environment in higher education today: the acceptable use policy and the data governance policy. While these policies may address all types of IT use on campus, they are most familiarly associated with the administrative (e.g., non-teaching or research) uses of IT resources and data on campus. Each policy addresses a different type of activity that may be influenced by IoT growth.

Acceptable Use Policies. Colleges and universities use acceptable use policies (AUPs) to tell students, faculty, staff and affiliated third parties how to use institutional IT resources. Like other industry sectors, these policies are used to help institutions protect IT resources because these resources are valuable business assets. These policies often contain a prescribed list of permitted and prohibited (i.e., "acceptable") uses of IT assets. People who violate the listed acceptable uses can face consequences that range from a reprimand, to restricted access to IT resources, to termination of employment. AUPs are considered a very important policy from a human resource and legal perspective because they promote workplace productivity and set forth an institution's intent to comply with regulatory requirements.

Higher education acceptable use policies not only have to state the limits of institutional IT resource use but must also serve as flexible codes of conduct that anticipate technology innovations. One such technology innovation is IoT and the rise of mobile device use on campus. Students, faculty, and staff use personally owned mobile devices every day to access the Internet, campus IT resources, and campus data. Campus AUPs often contain specific terms related to mobile device use in order to counteract the institutional security threats posed by those devices, namely the storage of institutional data on personally owned mobile devices and the fact that they are portable and easily lost or stolen.

Data Governance Policies. Data governance policies are a companion policy to the acceptable use policy. Instead of focusing on how a physical IT resource is used, however, these policies focus on defining institutional data and setting limits on how data are used. Data governance policies recognize that data are some of the most valuable assets that any college or university owns and they specify how students, faculty, and staff can use institutional data in different circumstances.

Often these policies specify certain roles with respect to data use. Two of the most important roles are data owners and data stewards or custodians. A *data owner* is a department or official who makes decisions about how that data are classified and how they may be used across the institution. For example, data that originates from an institution's financial aid office might be owned by an institution's bursar; data about student athletes might be owned by an athletic director, data about how an institution's IT infrastructure is set up might be owned by an institution's chief information officer. A *data steward or custodian* is a person who handles or uses data on a daily basis to accomplish a specified task, but is not a decision-maker for purposes of specifying a data classification level or permitted uses of data.

Data governance policies also often classify data into different levels based on the perceived importance or assumed sensitivity of the data. Generally speaking, data that are protected by law or regulation are classified at the highest level of sensitivity, while data that are generally publicly available are of a lesser level of sensitivity. These classification levels are used as the basis for developing data handling and use requirements for students, faculty, and staff to follow. These data handling requirements often address how users view, use, update, delete, or destroy data of different classification levels. It may be perfectly acceptable for a staff member to post an institution's course catalog on a public webpage, as that information is not particularly sensitive and a wide-range of campus members have a need to access course catalog information. That same staff member, however, would be prohibited from publishing student grades on a public web page because that data is highly sensitive and confidential, and protected by federal law.[1]

At most campuses, acceptable use policies and data governance policies work together due to a unifying underlying assumption: that a higher education institution can specify how institutionally owned IT resources are used (via its AUP) and how institutionally owned data must be handled (via a data governance policy). This assumption worked well enough in the BYOE environment because it also presumed that personal use of institutionally owned IT resources was incidental and *de minimis*. In the IoT environment this assumption is challenged. The cumulative effect of IoT devices on campus networks may no longer be incidental or *de minimis*. It also stands to reason that an institution could have legal obligations that arise as the result of personal data stored on institutionally owned IT resources. As important as they are now, acceptable use policies and data governance policies will become even more important to campuses in the IoT-enhanced future.

5.4.2 IoT in the Academy

IT use is pervasive across the non-administrative portions of colleges and universities as well. In the academy, there are two major intersections between information technology use in general and the IoT: teaching and research. In the classroom, IT policies are most often imposed by the faculty member or instructor running the class. For research functions, IT policies may be prescribed by the institution, the research funder, and even relevant laws and regulations.

Classroom Policies. The interplay between the rise of connected technology and its role in the classroom is complex. In some instances, connected technologies can be extremely useful for student note-taking and interacting with learning materials. Yet technology's use in the classroom can also be a distraction to student and faculty member alike. This leads to growing conflict between students and faculty over the appropriate use of technology in the classroom. In the classroom, faculty set the applicable use policy, and each faculty member may have a different policy for different devices. Few students report that institutional faculty encourage the use of mobile devices in the classroom (see Fig. 5.2). Faculty most frequently ban or discourage smartphone use in the classroom [2].

---

[1] It is outside of the scope of this chapter to talk about the different types of federal laws that apply to the data that colleges and universities use each day. One of the most common federal data laws referenced in the U.S. higher education environment is the Family Educational Rights and Privacy Act of 1974 (FERPA), which protects students and their families by ensuring the privacy of student educational records.

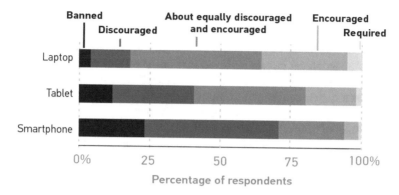

Fig. 5.2 Students' in-class experiences with their devices. *Figure by Kate Roesch, reprinted with permission from EDUCAUSE*

Banning technology in the classroom can hamper student success—47% of students rate smartphones as very/extremely important to academic success (among students who use smartphones in at least one course) [2]. From a policy level, whether student technology use in the classroom serves as a useful learning tool or distraction is left in the hands of faculty, although some institutions are considering policy at the institutional level [13].

While device proliferation and use of online learning options transform the modern student experience, so too will the IoT. One area in which the IoT will play a significant role is in the significant amounts of data that can be collected from personal devices and combined with institutionally collected data (from learning management systems and student information systems) to help create a clear picture of the student experience at the individual level. In particular this blended data can be used to inform everything from degree planning activities to predictive learning analytics systems. Learning analytics systems, which use data, analysis, and modeling to improve teaching and learning, can help students and institutions alike understand a student's past and present academic performance, and they can also be used to predict future student performance. These tools bring with them the security and privacy concerns already discussed earlier in this chapter. In addition, ethics questions about determinism (will the use of tools that predict future success in a certain course of study eliminate a student's free will to pursue their passions) and the extension of the institution in loco parentis (acting as a parent toward the student) arise.

Research Data Management Policies. Research data management policies provide guidance to all university constituents on how research data must be handled. These types of policies often look very similar to data governance policies, but they are specifically directed at the use of research data.[2] Entities that fund research, from U.S. federal entities like the National Science Foundation to private charities

---

[2] Research data is more than just scientific data. It also includes social sciences data, humanities texts, and any other data, regardless of form, produced in the pursuit of academic research.

or foundations, may also have research data management guidelines that institutions and researchers must follow. International laws, such as the European Union's General Data Protection Regulation, may also complicate the legal landscape by guaranteeing significant privacy rights to European research subjects (in the event that research involves human subjects). As research data is valuable intellectual property for a higher education institution and its faculty and researchers, making sure that data is properly protected is of utmost importance.

The Internet of Things adds new complications to research data management for higher education. Many IoT devices are wearable consumer devices that may provide especially useful data (e.g., habits, behavior, and location) that could be used in research to understand, treat, and prevent disease. Depending on how a researcher gathers this data, it could be considered human subjects research. Human subject research is research involving a living individual, about whom a researcher obtains data through intervention or interaction with the individual or identifiable private information [14]. In the United States there is a complex regulatory environment surrounding human subjects research which was designed to make sure such research is conducted in ethical ways. Existing relationships between consumer-provider and researcher-human subject are now possibly interwoven.

A Word About Cybersecurity Hygiene. Many institutions believe that they have a heightened role to educate students, faculty, and staff about cybersecurity hygiene. For faculty and staff, cybersecurity hygiene awareness and education activities are often mandated and part of an institutional compliance program. For students, cybersecurity hygiene education might be part of a general set of "self-defense" training provided to students as they enter college [15]. Education is in the institution's genes and existing tools and pedagogy can be leveraged to ensure that end users are properly aware of the risks and rewards presented by the IoT.

5.5 Next Steps in Campus IoT

Despite their sophistication, traditional campus policies, practices, and technology approaches may not provide much help unraveling the following scenario, which is probably fairly common today:

> *Emily uses her personal fitness tracker to track her health and the number of steps she takes each day. She has linked her fitness tracker data to her mobile health record from her medical provider so that her medical provider can monitor her fitness. Emily volunteers to participate in a campus research study, in which her heart rate is transmitted from the fitness tracker to a researcher daily. She accesses the web-based versions of both apps regularly from her personally owned mobile phone and from the campus computer lab. In the past, Emily has downloaded data from both her fitness tracker and her health record to view it on a campus lab PC. The data that she has downloaded is personally identifiable to Emily and contains medical diagnostic information as well.*

The scenario presents a number of questions that institutions may grapple with:

- What happens if there is a data breach:
  - On campus and Emily's medical data is exfiltrated from the campus PC?
  - On campus and Emily's medical data is exfiltrated from research storage?
  - At the personal fitness tracker provider's cloud storage?

- What happens if Emily's identifiable data is intercepted while transmitting from the device, via the campus network, to the device provider's cloud storage?
- Would the institution have a data breach reporting obligation for the exfiltration of Emily's data that is personally identifiable in any way?
- Does the institutional research data management policy articulate responsibilities and ownership for blended data?
- What if all the data, tied with her learning analytics data, reveals patterns of depression, suicide, or self-harm? Is there any responsibility to take action? On whom does this responsibility rest?
- Does the analysis change if Emily is a student (or alumna), faculty member, staff member, or member of the general public?

There are no hard and fast answers for higher education in the above scenario and the questions posed. Some of the answers to these questions will depend on laws and standards that require modernization and reframing. Some of the answers will depend on organizational culture, which is constantly evolving. Some of the answers will depend on the IoT market, some on consumer demands. There are many unknowns. As a Magic 8-ball might predict, the future is unclear.

The higher educational institution presents a microcosm of the Internet and society at large; an environment of elevated technology deployment, and valuable experience in utilizing and protecting personal and organizational information technology resources. While we had hoped to provide an overview of the higher education IoT environment with clear and concise IoT future-facing guidance to the readers based on the higher educational experience with BYOE, our drafting revealed that it may be too early to do so. In fact, our writing started to produce more questions than answers, leading us to the conclusion that neither higher education specifically, nor society at large, is truly ready for massive change that IoT growth will bring. To begin preparations for the future of IoT, higher education must understand the risk environment, undertake tremendous resource capacity planning, and modify its approach to technology policy.

Understanding the risk environment. The proliferation of IoT devices on campus poses both challenges and opportunities for higher education. A thorough understanding of current and future technology risks need to be explored and assessed regularly to determine and prioritize appropriate safeguards. IT departments will need to increase efforts in continuous monitoring to discover, identify, and analyze malicious activity quickly. Savvy security engineers will then use that newly formed analysis, constantly tuning security management systems to block threats. IT organizations will increase awareness of the external threat landscape, identifying signatures of malicious activity that can be thwarted at the institution's network borders.

At the same time, common usage behaviors within the campus network will be quickly recognized so that aberrant transmissions and connections can be questioned and stopped.

The extensive privacy risks introduced or exacerbated by IoT require new models of technical mitigation. Many institutions already implement safeguards that can help to mitigate these risks, but mitigation can be costly. To contain costs, institutions may currently limit the scope of protection to the most critical IT resources, data, and users. Perhaps faculty or staff must use multifactor authentication to access the student system whereas students access the system (only their own records) with just a password. Protected Health Information (PHI) may be contained on a restricted network subnet but student records are not. With the potential increase to the scale of the threat platform through IoT expansion, this prioritization methodology will likely require additional oversight and safeguards such as increased awareness and training and more frequent automated scanning for data leaks, perhaps even harsher disciplinary steps taken to curb risky activities.

Planning for Increased Resource Capacity. Resource capacity planning, especially for network throughput and speed, will continue to be a high priority for higher educational institutions, with the IoT ramping up the outcry for "MORE BANDWIDTH!"[3] To ensure mission-critical traffic flows through all the noise generated by IoT devices, institutions may find a need for increased network traffic prioritization, identifying traffic by type and throttling less-critical transmissions. IT organizations may also focus on developing and strictly adhering to termination policies and plans for outdated and underutilized technologies. With the onslaught of new system and network activity emanating from IoT devices, maintaining older systems becomes costly and risky.

Adapting Technology Policies. In addition to risks to security, privacy, and campus infrastructure, significant policy issues must also be considered. For example, can the institution set acceptable use limits on devices that are owned by end-users, and not the institution? What are the limits of data ownership and use when institutional and personal data are blended on institutionally and personally owned devices? Should mobile device use be integrated into the classroom? If so, how should technology-enabled learning and working spaces be configured, and how should any data gathered from any of these devices be used to improve pedagogy or student outcomes? What is the institution's responsibility for teaching students how to use connected technology properly and in a productive manner? Like other organizations, higher education institutions will need to be more future-facing in their approach to policy-making to try to address some of these questions.

While higher education's early experience with BYOE does put it ahead of the curve for analyzing Internet of Things opportunities and impacts, uncertainties remain. It is unlikely that a campus would ever tell its community members to leave their connected devices (tea kettles?) at the door. Nonetheless, do colleges and universities have the requisite fortitude to make sure that those connected devices are used in positive and enhancing ways? How does that change if the devices are provided by the institution?

---

[3] *MOAR* Bandwidth would be more appropriate! See e.g., https://imgflip.com/i/uug8z.

Picture a college or university, a bustling urban campus with modern buildings or a bucolic campus featuring an abundance of limestone and red brick. Imagine the data from hundreds of thousands of devices travelling across its expansive network harmoniously. Students, faculty, and staff enjoy the convenience of the devices and knowingly choose when and how to share their sensitive information. And the IT support professionals are sleeping well in their homes, comfortable in their knowledge that the network has ample capacity and appropriate safeguards to manage and protect devices and data.

References

1. Dahlstrom E, diFilipo S (2013) The consumerization of technology and the bring-your-own-everything (BYOE) era of higher education [Internet]. EDUCAUSE Center for Applied Research, Louisville, CO. [cited 25 Sept 2018]. Available from: https://www.educause.edu/ecar
2. Brooks D, Pomerantz J (2017) ECAR study of undergraduate students and information technology, 2017 [Internet]. EDUCAUSE Center for Analysis and Research, Louisville, CO. [cited 25 Sept 2018]. Available from: https://www.educause.edu/ecar
3. Tate E (2018) 2,300 Echo Dots put Alexa skills in every Saint Louis University dorm. EdScoop [Internet]. [cited 25 Sept 2018]. Available from: https://edscoop.com/saint-louis-university-amazon-alexa-echo-dot
4. Gartner (2017) Leading the IoT: Gartner insights on how to lead in a connected world, 2017 [Internet]. Gartner, Stamford, CT. [cited 25 Sept 2018]. Available from: https://www.gartner.com/imagesrv/books/iot/iotEbook_digital.pdf
5. Wheeler T (2013) The general data protection regulation sets privacy by default. [cited 2018 Sept 25]. In: Brookings Techtank Blog [Internet]. The Brookings Institution, Washington, DC. Available from: https://www.brookings.edu/blog/techtank/2018/05/23/the-general-data-protection-regulation-sets-privacy-by-default/
6. Grama J, Brooks D (2018) 2018 trends and technologies: domain reports [Internet]. EDUCAUSE Center for Analysis and Research, Louisville, CO. [cited 25 Sept 2018]. Available from: https://www.educause.edu/ecar
7. Sprint T (2018) In wake of "biggest ever" DDOS attack, experts say brace for more. [cited 25 Sept 2018]. In: Threat Post Blog [Internet]. Threatpost, Woburn, MA. Available from: https://threatpost.com/in-wake-of-biggest-ever-ddos-attack-experts-say-brace-for-more/130205/
8. Chronology of Data Breaches [Internet]. Privacy Rights Clearinghouse, San Diego, CA. [cited 2018 May 25]. Available from: https://www.privacyrights.org/data-breaches
9. Statista (2018) Internet of Things (IoT) connected devices installed base worldwide from 2015 to 2025 (in billions) [Internet]. Statista, Hamburg, Germany. [cited 25 May 2018]. Available from: https://www.statista.com/statistics/471264/iot-number-of-connected-devices-worldwide/
10. Addady M (2016) This university is making its students wear Fitbits. Fortune [Internet]. [cited 25 Sept 2018]. Available from: http://fortune.com/2016/02/04/christian-college-fitbit/
11. Yao M (2017) Your electronic medical records could be worth $1000 to hackers. Forbes [Internet]. [cited 25 Sept 2018]. Available from: https://www.forbes.com/sites/mariyayao/2017/04/14/your-electronic-medical-records-can-be-worth-1000-to-hackers/#776d5e1f50cf
12. Pomerantz J, Brooks D (2017) ECAR study of faculty and information technology, 2017 [Internet]. EDUCAUSE Center for Analysis and Research, Louisville, CO. [cited 25 Sept 2018]. Available from: https://www.educause.edu/ecar

13. Fernandez L (2018) Should IT provide faculty with tools to disable Wi-Fi in their classrooms? [cited 25 Sept 2018]. In: Transforming Higher Ed Blog [Internet]. EDUCAUSE Review, Louisville, CO. Available from: https://er.educause.edu/blogs/2018/4/should-it-provide-faculty-with-tools-to-disable-wi-fi-in-their-classrooms
14. Protection of Human Subjects, 45 C.F.R. Sect. 46.102 (2018)
15. Brooks D, Pomerantz J (2017) Beyond passwords and PINS: students and infosec training. [cited 25 Sept 2018]. In: Data Bytes Blog [Internet]. EDUCAUSE Review, Louisville, CO. Available from: https://er.educause.edu/blogs/2017/10/beyond-passwords-and-pins-students-and-infosec-training

Chapter 6
Trust, Identity, Privacy, and Security for a Smart Campus

Karen Herrington

6.1 Introduction to Identity Management

Who are you? This is a subconscious question that our brain asks and automatically answers many times as we encounter other people in our daily lives. How do we know who someone is? We might recognize the other person by sight, their physical appearance, or by the sound of their voice or perhaps by context—You are the mailman. I always see you delivering my mail.

But what about interactions that take place in the virtual or cyber world? How do we know who we are interacting with? We cannot see or hear the other person. I am sure everyone has heard the old joke, "On the internet, no one knows you're a dog." We might require that the other person type their name as a means of identification. But how can we trust that the person typing "Jane Smith" really is Jane Smith? One way might be to require that the person enter a shared secret, or password, in addition to their name. We might even require that they enter more than one shared secret for additional assurance that they are who they say they are. This process of users of online applications providing names and passwords as a means of identification is called authentication, and secure authentication is foundational to the management of online identities.

Once a virtual person has established who they are, the next question to be answered is "What can you do?" Let us draw another parallel to the real world. My mailman is allowed to put my incoming mail into my mailbox and take my outgoing mail out of my mailbox. He is not allowed to come into my house or get into my car. In the cyber world, we need some method of tracking and enforcing what a virtual person is allowed to do. Let us say we want Jane Smith to be able to use our online service. So we allow Jane to create an account. She enters her name,

K. Herrington (✉)
Blacksburg, VA, USA
e-mail: kmherrin@vt.edu

address, and phone number, and saves this information as part of her profile. According to our rules, Jane will have full privileges to view and change any of her demographic information in the future. Perhaps we also allow Jane to specify an alternate account holder and she elects to have John Smith as her alternate. John might only be allowed to view information in the account but not change it. This process of tracking and enforcing a virtual person's privileges or permissions is called authorization. Authorization is another foundational component of online identity management.

6.2   Identity Management and the University Campus

University campuses are active, vibrant places in which many different types of people are associating and engaging every day. Large universities today can have more than 200,000 students, 100,000 employees and 1,000,000 living alumni. Not to mention countless parents, visiting scholars, researchers, and others that come and go as student cohorts, courses and projects change from semester to semester. All of these varied types of individuals expect to interact with the university virtually. According to Pew Research Center, 89% of all American adults are internet users, and among adults younger than 50 years old, the percentage increases to 97% [1].

So how does a university campus go about managing all of these online identities? Many begin by maintaining some sort of electronic person registry that contains current information about everyone that needs to interact with the university in a virtual manner. The registry information includes things like names, addresses, phone numbers, birthdates, affiliations, and identification numbers. Further, the affiliated individuals are allowed to create and save shared secrets or passwords to facilitate a continued virtual relationship with the university. In order to increase the assurance that the same virtual person is returning for each subsequent interaction, the institution will have strong password management processes in place. These processes might include dictating the length and the complexity of passwords—that is, the password must contain upper case and lower case letters, numbers and special characters. The rules might also include specifying the length of time that a particular password may be used before it must be changed or requiring the use of more than one of the following types of factors—something you know, something you have, and something you are—commonly called multifactor authentication. An example of multifactor authentication would be entering a password (something you know) as well as entering a code that has been sent to your cellphone (something you have) as further confirmation of your identity.

As mentioned, one element of the person registry is an individual's affiliations—how the person is associated with the university. For instance, a person might be a student or an employee or a researcher, or perhaps all three. The online services that an individual is allowed to access will be based partially on their affiliations. Only students can register for classes. Only employees can view a record of their job earnings. Only a researcher can access the files or experiment results associated with a particular project. The process of granting access to services for which an individual is entitled is called provisioning.

A person's affiliations will typically change over time. A student will graduate and become an alumnus. An employee might leave the university for employment elsewhere or perhaps retire from the university. A researcher's project will end. The converse process to provisioning is deprovisioning, or removal of access when one is no longer entitled. Timely and appropriate deprovisioning of access is an important security function of the institution.

6.3   Identities in the Internet of Things (IoT) Ecosystem

The fundamental questions "Who are you?" and "What can you do?" become infinitely more complex in the world of the Internet of Things. Whereas traditional identity management has been concerned primarily with people interacting with online services, the ecosystem of IoT includes devices, people, software, firmware, and the Cloud. Relationships can exist between any of these "actors," and interactions involving authentication and authorization must be orchestrated and managed in multiple scenarios. Let us examine this interplay between identities and the Internet of Things more closely (Fig. 6.1).

The life cycle of a device begins when the device is manufactured and shipped. Once the device arrives at its destination, it must be registered into the environment in which it will operate so that its identity will be known and recognized by other actors in the environment. By definition, IoT devices are intended to connect to a network to transfer collected data and receive commands for actions. These interchanges of data and commands dictate that the device must have a valid credential

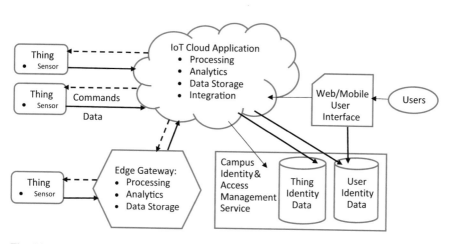

Fig. 6.1 IoT Identity Management Reference Architecture

to access the network, as well as valid credentials to deposit data in data stores or perhaps access other downstream devices or applications in order to complete actions. In addition to authentication—proving that you are who you say you are—the messages that travel to and from the device over the network, must carry information about what the "Thing" is allowed to do—its authorizations. The credentials for authentication and the authorized privileges may need to change over time and ultimately, the device will need to be deprovisioned when it reaches the end of its useful life cycle.

How do devices acquire their authentication credentials and authorization privileges, and how are they managed? The process begins when the device arrives at its destination and a person must prepare the device to perform in the environment in which it will operate. This might be an engineer or operator setting up a piece of equipment in an organizational setting or it might be an individual with a personally owned appliance or apparatus. By virtue of their job responsibilities in an organization or right of personal ownership, the individual is empowered to create or change the authentication credentials of the device and specify the actions that the device is allowed to perform. Thus, another realm of interaction and life cycle has been established and must be managed—that of the relationship between the individual and the device. An employee who is responsible for a device or set of devices might change positions within the organization or terminate employment. An individual who owns an appliance might sell or give it to someone else. In each case, the device privileges of the original administrator will need to be deprovisioned and the new administrator will need to be provisioned. In some cases, multiple people might have been given access privileges on the device, such as the situation in a home setting where several family members are able to reprogram the thermostat or smart television.

6.4   IoT Identity Management and the University Campus

We have looked at how identities are typically managed on a university campus and some of the processes that are in place to ensure that the university can perform effectively. Let us look at how the Internet of Things challenges that traditional model of identity management.

Recall that traditional campus identity management is mainly concerned with people identities and utilizes a person registry to record demographic information and credentials associated with the identities within its sphere. When IoT is added to the mix, devices must be registered into the environment as well. However, "Things" have different characteristics or "demographic" information that does not translate well into the traditional model of a person registry. This will necessitate the use of a more generalized entity registry that is better suited to the types of information that must be recorded and managed about a "Thing."

Customary provisioning and deprovisioning practices will be severely impacted. The majority of provisioning and deprovisioning of identities on a university campus tends to follow the ebb and flow of the instructional cycles of the university. Each fall, it is expected that a whole new crop of students and faculty members will become members of the university community. In reality, the admissions processes for students and hiring processes for employees are initiated months and sometimes almost a year before these individuals actually arrive on the campus. It is through these static, well-established, lengthy processes that demographic information is gathered, identities are vetted and credentials are issued so that students and faculty are already provisioned when the semester begins.

Association of IoT devices into the university environment will take place in a much more dynamic and unpredictable manner. Gartner predicts that there will be nearly 21 billion Internet of Things (IoT) devices deployed by 2020 [2] and some predictions are even higher. This ubiquitous nature implies that devices could randomly appear on or disappear from the network at any given time. It also suggests that an institution's provisioning and deprovisioning processes will need to become much simpler, quicker, and more robust to accommodate the potential churn, while also maintaining the security of the campus network and systems. Today, there are transaction-based termination processes in Human Resources that trigger the removal of a departing employee's access privileges, as well as Registrar processes that realign student privileges as students transition to alumni. What will trigger the deprovisioning of a device? How will the identity management system know that the device is "finished" and will not appear again? This will be akin to the uncertainty regarding the status of a student who stops attending classes in the middle of the semester but fails to withdraw from the university or completes the spring semester as a junior but fails to return in the fall for senior-year classes. However, the uncertainty will be compounded as the appearance and disappearance of devices from the campus will be much swifter, higher volume, and circular in nature.

The multiplicity of relationships that are possible and necessary in the world of the Internet of Things will be difficult to represent with prevailing higher education identity protocols. A "Thing" might have one set of permissions when acting on behalf of a particular user and another set of permissions when acting on behalf of a different user. Likewise, multiple software applications might need to interact with a device—each conveying a different set of actions. Permissions might need to be shared among several users of devices or delegated from one user to another user for a limited period of time and then revoked.

Many IoT devices utilize transport protocols that are optimized for use in situations where a "small code footprint" is required or the network bandwidth is limited. Reduced or nonexistent input/output capabilities and the use of constrained protocols will severely inhibit the campus identity infrastructure's ability to interact and communicate with these devices. Higher education identity protocols, in use today, will need to be reexamined and adapted to integrate with common IoT protocols and transfer appropriate and complete authentication and authorization information. In situations where devices utilize passwords for authentication, the limitations of the devices may make it impossible to adhere to strong security practices such as enforcing password complexity and expiration rules or requiring the use of multifactor authentication.

Finally, the type of IoT devices known as edge devices will provide a unique challenge to a comprehensive identity management strategy. Edge devices are a particular class of IoT devices that sit at the edge of the network and are powerful enough to run full-fledged operating systems and complex algorithms. They are capable of doing computations locally and can execute self-contained loops of processing data and generating commands without having to "call-back" to central-ized applications [3]. Since these devices operate somewhat independently, how will they be incorporated into the central identity management system? How will the devices be provisioned and deprovisioned if the central identity management system does not know about them? How can authentication and authorization take place?

6.5   The Smart Campus

University campuses are no stranger to the Internet of Things. Carnegie Mellon University is widely credited with the creation of the first known IoT device in 1982. Michael Kazar, former Carnegie Mellon graduate student: "It really was not a very significant thing at the time. There was a Coke machine on the third floor of this eight-story building, and people didn't like the fact that they would go down all the way to the third floor and discover that the Coke machine was empty. Someone said, "Hey, why don't we set it all up so the Coke machine was on the Internet." The way the thing was structured is you had the Coke machine and then a serial line connecting the Coke machine to some terminal concentrator that it just so happened we had control over the source code. It could check ten times a second" [4, 5].

Then, in 1993, students at Cambridge University were motivated to Internet-enable a coffeepot. Paul Jardetzky, former Cambridge graduate student: "We were inspired by the fact that none of us could get a cup of coffee because we had one coffeepot for multiple floors in a seven-story tower. We had a graphics group that had some cameras lying around, and we had frame grabbers lying around, and we hooked it all up and we put it in the room with the coffeepot. When we put it on the Web, there were so many hits that it had gone what we called viral today" [5, 6].

From those early beginnings of experimenting with novelty, Internet-connected Coke machines and coffeepots, universities are starting to recognize the true value of a Smart Campus—a campus that links devices, applications, and people to enable new experiences and services and improve operational efficiency. At Sun Devil Stadium on the campus of Arizona State University (ASU) in Tempe, sensors con-nected to the WiFi and cellular network collect temperature, humidity, and noise data for use by facilities staff. As part of a longstanding cheering contest, the noise data analysis identifies the section of the stadium that is making the most noise and puts the results on a big screen. Sensors can identify if a faucet anywhere in the stadium is left running after a football game is over, to help cut water usage. ASU is also exploring providing information through a mobile app on the availability of parking and wait time estimations for concession lines and restrooms. The tech-infused stadium has been a test bed for a larger investigation of and investment in Internet of Things technologies at ASU, according to Gordon Wishon, former university chief information officer. Wishon believes that research universities provide the perfect place to test and deploy IoT. "The enterprise of a large research university has some component of every industry vertical in the larger world around us. We not only support academic and research operations, but also very large business enterprises with retail operations, transportation, healthcare, ticketing, supply chain," he said [7].

In March of 2017, Georgia Tech's Campus Recreation Complex (CRC) launched a pilot version of a health and wellness platform, created by Cytilife [8], to their student body. The goal was to help improve Georgia Tech students' overall health by making the CRC an integral part of the student's everyday life, one where they both use the CRC and enjoy the experience. The platform takes into account a student's schedule, their personal gym time preferences, as well as how crowded the gym is, and uses these data points to inform the student about what time would be the best for them to go to the gym. The platform also has the ability to help the students plan with their friends a good time to go to the gym and may also have the ability in the future to let the students know if there is parking available near the CRC or the best way to get to the CRC without delays [9].

6.6   Adapting Identity Management for IoT

Given the challenges and disparities between traditional identity management of people and the management of the identities of "Things", how will university campuses adapt? The customary person registry that institutions use to record demographic information about their Internet users will need to evolve into a more flexible, expanded entity registry that can accommodate relevant device information in addition to people information. Virginia Tech is one institution that is beginning this evolution with the creation of a system containing comprehensive information about physical facilities using the concepts of Building Information Modeling as a foundation. When completed, this facility information system will contain serial numbers and descriptions of devices, dates of installation, maintenance and update schedules, and other pertinent data for all devices in university buildings. As part of the effort, standardization is occurring around nomenclature, units of measurement, and the scope of data that needs to be collected as new devices are added to buildings. Employee roles such as designer, constructor, and maintainer are being defined with respect to responsibilities and authorizations. When assigned, these roles will establish relationships between people and devices that can be managed throughout the life cycle of "Things".

An identity ecosystem that potentially includes millions of "Things" and thousands of login or access validation actions per second will require flexible, extensible identity protocols and a responsive, scalable infrastructure that is able to represent, transport, and communicate complex identity interactions. Many believe that the OAuth2, Open ID Connect, and User-Managed Access (UMA) standards and tools coupled with a Representational State Transfer (REST) Application Program Interface (API) framework will enable the necessary capabilities [10, 11].

OAuth2 and Open ID Connect are token-based authentication and authorization standards and profiles. OAuth2 uses issued tokens to enable delegated access to server resources on behalf of a resource owner and could allow a device to gain the necessary permissions to represent a user to internal and external services. The device permissions could be quickly and easily revoked by simply invalidating the assigned OAuth2 access tokens. Privacy tools like UMA could provide needed granular consent capabilities by enabling control over what devices, services, and users can access data, for how long, and under what conditions [10, 11].

APIs allow two software programs to communicate with each other by providing the complete set of rules and specifications to follow in order to facilitate the interaction. A REST API framework adheres to several constraints that make it attractive for connecting users, devices and things to applications and services in an IoT environment. Statelessness, meaning that each call is independent and contains all of the data necessary to complete itself successfully, and a uniform interface that allows for communication in a single language, independent of the architectural backend of either side, provide the flexibility needed in an ecosystem consisting of a range of services and applications on many different platforms and languages. The REST API constraint of layered systems, with each layer having a specific functionality and responsibility and different layers of the architecture working together to build a hierarchy, can provide the necessary abstraction and modularity for a scalable IoT environment [12].

Authentication processes, such as enforcing password complexity and expiration, are cornerstones of today's campus identity management strategy. However, these types of security practices will be insufficient in the complex IoT world of users, devices, things, services, and applications with multiple and varied relationships. Context-based and continual security have been suggested as critical approaches to securing IoT. Context-based security is more robust than username/password authentication and could use indicators like geographic location, Internet Protocol (IP) address, time of day, and device profile to generate a real-time risk score of the confidence level that an "actor" is who they say they are. Couple this with continual security—calculating and checking the risk score multiple times during a digital session instead of only at initial authentication—and this could ensure the authenticity of "Things" at all times and spawn actions to mitigate the risk whenever an anomaly is detected. In the world of IoT, universities must prepare to have mature authentication processes that extend beyond passwords to include biometrics, certificates, and other forms of password-less authentication [10, 11].

The Internet of Things holds promise for making our university campuses more efficient, sustainable, and enjoyable places for students to live and learn. However, to date, identity and access management in IoT has largely been an afterthought or not considered at all. Secure, comprehensive IoT identity management strategies and solutions that enable persistent identity across all touchpoints and interactions will be essential in enabling higher education institutions to realize the vision of the campus of the future—the Smart Campus.

References

1. Pew Research Center: Internet & Technology. Internet/Broadband Fact Sheet: Who uses the internet. 2018 Feb 5. Available from: http://www.pewinternet.org/fact-sheet/internet-broadband/
2. Gartner, Inc. Newsroom Press Release: Gartner Says 8.4 Billion Connected "Things" Will Be in Use in 2017, Up 31 Percent From 2016. 2017 Feb 7. Available from: https://www.gartner.com/en/newsroom/press-releases/2017-02-07-gartner-says-8-billion-connected-things-will-be-in-use-in-2017-up-31-percent-from-2016
3. Reale A (2017) A guide to Edge IoT analytics. Available from: https://www.ibm.com/blogs/internet-of-things/edge-iot-analytics/
4. Teicher J (2018) The little-known story of the first IoT device. Available from: https://www.ibm.com/blogs/industries/little-known-story-first-iot-device/
5. Vinik D (2015) Politico magazine. The Internet of Things: an oral history. Available from: https://www.politico.com/agenda/story/2015/06/history-of-internet-of-things-000104#
6. Kesby R. (2012) BBC News: how the world's first webcam made a coffee pot famous. Available from: https://www.bbc.com/news/technology-20439301
7. Raths D (2017) Campus Technology: 'Smart' Campuses Invest in the Internet of Things. Available from: https://campustechnology.com/articles/2017/08/24/smart-campuses-invest-in-the-internet-of-things.aspx
8. Cytilife, Inc. Available from: https://www.f6s.com/cytilifeinc
9. Georgia Institute of Technology. Smart Campus Case Study: The Connected Campus: Using Sports Technology to Improve Student Health, Well-being, and Success. 2017 August
10. FORGEROCK. Your Guide to the identity of things: the top 12 considerations for an IoT ready identity and access management platform. Available from: https://www.forgerock.com/resources/view/64142260/e-book/your-guide-to-the-identity-of-things.pdf
11. FORGEROCK (2018) A reference architecture for the Internet of Things. Available from: https://www.forgerock.com/resources/view/64295616/overview/identity-of-things-a-reference-architecture.pdf
12. Mulesoft, LLC (2019) What is a REST API? Available from: https://www.mulesoft.com/resources/api/what-is-rest-api-design

Chapter 7
Security for Science: How One Thing Leads to Another

Hannah Short

7.1 Introduction

I have a vivid memory as a child of a particularly austere English teacher berating me for overuse of the word "Thing". Imagine my relief as we entered the 2010s, "Thing" found its way into the spotlight and I was able to freely and legitimately reinsert it into my vocabulary! In my childhood, "Thing" was criticised for being imprecise. I enjoyed many conversations with friends and colleagues whilst researching this chapter, and ventured to ask for their own definition of a "Thing" from the Internet of Things (IoT). The answers I received have led me to believe that my English teacher was, in fact, ahead of her time; outside a small circle of experts the concept of "Thing" is indeed imprecise. This chapter will exploit this imprecision as we discuss the overlap between the IoT and Science. We will touch on the experiments, laboratories and scientists impacted, plus the "Things" themselves.

As a member of the Computer Security team at the European Council for Nuclear Research (CERN), the content here has a declared bias towards High Energy Physics. Many of the ideas, however, are relevant further afield.

7.1.1 The Coolest, Largest, and Fastest Things on Earth

Scientific experiments are designed to push the limits. Some of the most interesting hypotheses currently being tested focus on questions that require powerful, specialised and often complex machines to be constructed. The Square Kilometre Array will be the largest observatory ever built; with a total collecting area of well

H. Short (✉)
Ferney Voltaire, France
e-mail: Hannah.short@cern.ch

© Springer Nature Switzerland AG 2019
F. D. Hudson (ed.), *Women Securing the Future with TIPPSS for IoT*, Women in Engineering and Science, https://doi.org/10.1007/978-3-030-15705-0_7

over one square kilometre (or one million square metres) spread over two continents [1]. Researchers at the National Institute of Standards and Technology (NIST) are using lasers to create temperatures colder than the coldest regions of the universe [2]. There was great excitement when it seemed plausible that particles had travelled faster than the speed of light between CERN and Gran Sasso in Italy, a distance of over 700 km through the crust of the earth [3]. This turned out to be a false alarm but the principle is the same. We know how the rules of science work in our environment, to make discoveries we need to probe the extremes.

Experiments can span countries, continents and occasionally planets. They rely on physical networks (and sometimes more inventive solutions involving satellites, etc.) to transfer data. Attached to these networks are the actors involved in our workflows; the experiments that produce the data, the machines that process and store the outputs, the researchers that perform their analyses. Each actor introduces risks into scientific workflows.

There are several concrete ways in which IoT devices have been recognised to play a part in scientific workflows and experiments today:

- IoT devices integrated with the control systems of experiments, such as sensors.
- Connected scientific apparatus in the laboratory, such as thermometers and oscilloscopes.
- "Custom" IoT devices developed by researchers.

The experiment control systems, the software and hardware that configure the machinery and electronics at the heart of an experiment, will typically not be connected to the internet directly but separated on a dedicated network. At CERN, this is called the Technical Network, shown in Fig. 7.1, and offers near complete isolation from the outside world. Devices may be connected at any point in the chain, both on the Technical Network and General Purpose network that is connected to the Internet. IoT devices on a Technical Network, whilst not necessarily conforming to the "internet" condition of being an "Internet of Things" device, equally pose security concerns for science and this chapter includes all networked IoT-like devices.

It is essential that a defence-in-depth approach is taken towards IoT security for science, with best practices incorporated in network hygiene, security awareness campaigns and procurement practices, to name but a few relevant aspects. This chapter takes a closer look at security for distributed science, IoT in a laboratory setting, and focuses on an example taken from CERN's recent campaign to identify and secure connected devices.

7.1.2   Science as a Target

Research, as any other sector, has its own particular threats. Two key risks, as highlighted in SURF's Cyber Threat Assessment 2017, are that of "obtaining and publicizing data" and "espionage" [5]. It may seem perverse that research, particularly

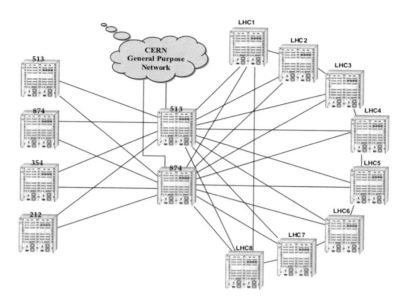

Fig. 7.1 Technical Network schematic, from the Large Hadron Collider (LHC) Design Report, 2004. The LHC is CERN's particle accelerator that sits 100 m below the ground and measures approximately 27 km in circumference. End node security and updates are paramount; "It must be noted that because the technical network infrastructure is interconnected with the general purpose network, security break-ins can be attempted on the devices connected to this infrastructure. End nodes security survey and updates must not be forgotten" [4]

open research initiatives, should be the target of espionage. The science will be publicly available, so why steal it? Until the point that science is published, there is usually a need to ensure confidentiality of data. Two key reasons for this are firstly that those paying for time on an experiment may be entitled to exclusive access during a certain period, and secondly that a level of separation of information between research groups is often necessary to avoid bias and ensure scientific integrity. The target audience (potential customers!) of stolen scientific data is small, meaning that organisational data that provides insights into new technologies, planned financial decisions of institutes or personal information of individuals may be a more interesting target for espionage.

Certain Research fields, particularly those dealing with topics in the popular domain such as nuclear research or genomics, are swathed in conspiracy theories that make them interesting targets for "hacktivists" (activist hackers) and others wishing to cause damage to reputation. For example, defacing a single research website may provoke a media storm culminating in a headline stating that a multimillion euro experiment was "almost hacked".

An additional target for online attackers is the abuse of computing power. The potential to exploit computing capabilities for disruption (e.g. denial of service attacks) or financial purposes (e.g. digital currency mining) can be an attractive incentive and Research organisations play home to powerful resources.

The threat landscape for Research Institutes is complex and includes threats typical to both Industry and the Education sectors. The inclusion of IoT devices into this landscape may both alter the existing attack vectors and introduce new opportunities that focus on Science as a target.

7.2 No Scientist Is an Island

7.2.1 Connected Communities, Data and People

The individuals working on an experiment are mobile and highly interconnected. Typically, laboratories will welcome visiting scientists who will bring with them their own devices (laptops, phones) and contribute under the affiliation to their home organisation. A scientist may spend 40 years working on an experiment but change university ten times in the process. Many research communities not only encourage researchers to work on their own devices but also to continue to use credentials (username and password, certificates, authorisation tokens, etc.) from their home organisation. For example, someone may contribute to the Laser Interferometer Gravitational Wave Observatory (LIGO) using a digital identity from the University of Edinburgh [6]. This identity could be used to allow the researcher to access one of LIGO's underlying computing centres based anywhere in the world and shared by other Research Communities. Security vulnerabilities can be introduced at the home organisation, by LIGO themselves, or at the underlying infrastructure, and can propagate throughout the stack. The mobility of researchers means that attack vectors exist to link organisations and institutes, which, at first glance, seem unrelated.

To cope with the connectedness of actors in scientific workflows, a strong layer of policy and trust frameworks is necessary to ensure that each organisation operates in line with a common baseline of acceptable operational security. This baseline may need to be updated as previous assumptions become invalid in the era of IoT. Whereas before we were dealing with PCs and portable devices that were well understood, homogeneous, and largely inaccessible from the outside, all bets are now off with the variety of devices available.

7.2.2 Joint Incident Response and Trust

There is no such thing as 100% security. Since you are reading this book, I assume this is a concept that you have already accepted so will not spend long trying to convince you. Security professionals are in a constant battle against attackers, with new vulnerabilities periodically emerging; as proactive as your security measures may be there will be a gap between a vulnerability's disclosure and your mitigating actions. Sometimes you will fall victim to an attack. Once we accept the inevitability of security incidents it becomes clear that a fundamental component of a robust security programme is Incident Response. Incident Response encompasses many phases, principally; preparation, containment, investigation, resolution and post-incident review of procedures and practices. Responsive collaboration of service and network operators, forensics experts, public communications and policy makers is critical.

Large-scale science tends to rely on distributed computing infrastructures, where resources are incorporated into a computing pool. Computing power may come from member organisations, supercomputing centres, and increasingly commercial infrastructure providers. In this model, an efficient response to a security incident becomes a problem of coordination and trust. Distributed Computing requires the collaboration of each participant that contributes to the computing needs of an experiment. Shared policies, procedures and the fostering of trusting relationships between participants are critical to ensuring that an incident can be successfully resolved [7]. Failure to build the required level of coordination will result in a suboptimal response. There are myriad ways in which the Incident Response process can be disrupted; some examples are loss of evidence (a contributing data centre reinstalls a system rather than gathers forensic evidence), inability to deploy defensive measures (a site fails to patch systems due to a poorly understood dependency) or a leak of confidential information to the media (without pre-established disclosure agreements or identification of a designated communication manager an individual may respond to journalists and cause damage to reputation). As with operational security, there are existing frameworks that identify the correct behaviour between computing providers [7] during Incident Response. It remains to be seen exactly how consideration of IoT might have an impact.

7.3 IoT in the Laboratory

7.3.1 Mitigating Curiosity

It is widely held that "curiosity killed the cat", but for a scientist curiosity is an essential ingredient of research. As a defining characteristic of many researchers, curiosity has the potential to introduce significant risks when it comes to security. Scientists and engineers tend to have both the technical knowledge and drive to be among the first to test out new technologies. Walk into a physicist's garage and you may well find an arduino geared up to measure the humidity and report back to its owner in case of excess moisture, step onto their balcony and their herbs may be watered autonomously by a similar setup. When we talk about IoT and Scientists, we are not only talking off-the-shelf IoT. Such homemade devices may be set up once (with a thought to secure configuration if we are lucky) and possibly never touched again.

This curiosity for technology is what has driven many students to research in the first place but applied to connected devices it has the potential to introduce unanticipated risks. A certain level of security education should be given to any scientist dealing with computing topics or infrastructure and, wherever possible, secure solutions and tools should be offered and maintained by laboratories and institutes. Security education, of course, stretches far beyond IoT security. Training in secure software development, access control, privacy fundamentals, to name a few areas, should be made available to scientists in the interest of the wider community.

7.3.2 Abundance of IoT

In 2016, CERN began a wide scale scan of its network to identify IoT devices [8]. You might ask yourself why a scan was required; surely it should be clear which devices are connected to a network? The principal reason is the following: bring-your-own device is standard at many Research Institutes. Granting network access to an abundance of visiting scientists, often in the thousands and increasingly with multiple devices, brings a certain overhead that, at scale, must be delegated to the scientists themselves who are trusted to register devices. Scientists are able to request access for their own devices and specify characteristics (operating system, vendor, etc.) manually. Since having total control over the devices connected to the network is unachievable, instead, effort is invested in policy, monitoring, incident response and network security. The introduction of IoT devices into society has been a slowly evolving process, with the result that CERN—like other laboratories and institutes—has devices connected to the network alongside desktop PCs, phones and laptops. It has become commonly acknowledged that good cyber hygiene recommends that IoT devices be connected to a network with tighter controls than those appropriate for laptops and desktop PCs. To be able to move such devices to a secure network architecture, the first step is to identify them.

A preliminary scan of the CERN network led to the identification of approximately 3000 devices distributed across the General Purpose Network and the Technical Network [9]. These devices were split into categories based on information gathered either through CERN's database of device information or through additional network analysis. The breakdown of devices can be seen in Fig. 7.2, with Routers and Switches, Webcams, Virtual Network Computing Viewers and Printers being the most abundant. In addition to these general-purpose devices, a number of scientific instruments were found. Notably thermometers.

Thermometers are used around CERN, sometimes as stand-alone apparatus and sometimes attached to experiments and integrated into their configuration systems. The ability to override a thermometer could directly impact the operation of a scientific experiment. I will let you use your imagination as to how, precisely, but to give one possibility—safeguards against overheating could be triggered by increasing the measured temperature leading to system shutdown, unavailability and loss of data taking. There are opportunities for holding devices to ransom, and direct financial consequences from the missed data capture. A more subtle concern

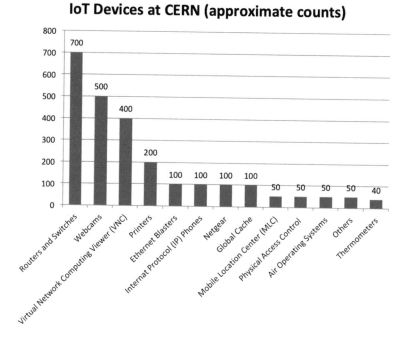

Fig. 7.2 Approximate numbers of IoT devices found at CERN, on both the General Purpose and Technical Networks [9]

is the potential for the integrity of data to be compromised and lead to invalid scientific results.

Whilst investigating the security vulnerabilities of Papouch TME Thermometers, commonly used in laboratories, a number of concerns were highlighted [9]:

- There was no authentication required for access to the Web User Interface or Telnet port.
- The Web User Interface allowed a range of actions, including changing the firmware and device configuration.
- A superadmin account was enabled by default, whose username and password cannot be overwritten.

All three of these aspects have the potential to cause significant disruption. In particular, the ability to arbitrarily change the code running on a device connected to a large-scale physics experiment should give cause for concern. In the case of CERN these vulnerabilities were either quickly addressed with the collaboration of the device owner, or the device was disconnected from the network. However, a one off scanning activity is not enough. IoT devices should be securely configured, facilitated by training for scientists and specific security audits, and their maintenance folded in to ongoing security processes. The impact of a compromised IoT device should be mitigated as an absolute priority. Trusting a single IoT device as part of a scientific workflow presents a particularly inviting vector of attack; the likelihood of undetected compromise can be significantly reduced by deploying multiple devices, and by calibrating against "un-connected" devices.

In a domain like physics the consequences can be great, but typically the experiments can be repaired and the data retaken. When research domains deal with living samples, the consequences can be more grave. Connected fridges and freezers that alert the researcher to unexpected temperature changes offer the opportunity to avert potential thawing but at the same time may introduce the possibility for a malicious actor to do just the opposite [10].

7.4   Where Are We Heading?

IoT devices are set to play a major role in scientific workflows. In the future they may offer considerable benefit to data taking and experiment configuration, with scientists already expressing interest in their potential. It is perhaps early days to make concrete predictions; the full impact of IoT is yet to be understood for the laboratories, the computing infrastructures and the individual researchers. However, there is no denying that IoT devices are already present and proactive measures should be taken to mitigate the risks that have been introduced:

- Network hygiene practices for Research Institutes must evolve to mitigate IoT risks. Appropriate network configuration should be the default and checks for known vulnerabilities should be made on a periodic basis. The message from the Large Hadron Collider Design Report, 2004, in Fig. 7.1 is still valid, and particularly pertinent for IoT: "End nodes security survey and updates must not be forgotten".
- Measures should be put in place to minimize the impact of an exploited device. IoT devices for scientific measurements should not be used without calibration against "un-connected" devices. Any deviation should be investigated. An additional safeguard may be to deploy multiple IoT devices and calibrate between them.
- Policies and procedures should evolve to include specific measures for IoT. This may include the augmentation of policies that span the multiple organisations and infrastructures that contribute to global science.
- Security training for scientists is required to highlight the risks of connected devices and the need to undertake certain security measures such as upgrading firmware and changing default passwords.

As, I am sure, will be mentioned in other chapters of this book, there is a strong hope that IoT vendors will step up and improve the security of their products. This is particularly important in scientific equipment such as cooling, heating and measurement devices where it is possible that compromise could lead to experiment malfunction and ultimately financial or physical risk.

Acknowledgments Pascal Oser (CERN) and Sharad Agarwal (CERN), for their work on scanning CERN's network for IoT devices. Stefan Lueders (CERN) and Romain Wartel (CERN) for insight into attack vectors against laboratories. A particular thank you to Clement Onime from The Abdus Salam International Centre for Theoretical Physics for the insightful discussion.

References

1. How The SKA Telescope Will Be Spread Out Across Two Continents, https://www.skatelescope.org/layout/; 2018. Available from: https://www.skatelescope.org/layout/
2. Clark JB, Lecocq F, Simmonds RW, Aumentado J, Teufel JD (2017) Sideband cooling beyond the quantum back action limit with squeezed light. Nature 541:191 EP. https://doi.org/10.1038/nature20604
3. ICARUS Collaboration, Antonello M, Aprili P, Baiboussinov B, Baldo Ceolin M, Benetti P et al (2012) Measurement of the neutrino velocity with the ICARUS detector at the CNGS beam. Phys Lett B 713:17–22
4. Brüning OS, Collier P, Lebrun P, Myers S, Ostojic R, Poole J et al (2004) LHC design report. CERN yellow reports: Monographs. CERN, Geneva. Available from: https://cds.cern.ch/record/782076
5. SURF Cyber Threat Assessment 2017, Education and Research Sectors (2018) Available from: https://www.surf.nl/binaries/content/assets/surf/en/knowledgebase/2017/surfcyberthreatassessment.pdf
6. Atherton CJ, Barton T, Basney J, Broeder D, Costa A, van Daalen M, et al (2018) Federated identity management for research collaborations. Available from: https://doi.org/10.5281/zenodo.1307551
7. Short H, Wartel R (2016) Building security and trust in inter-federation. Proc Sci. https://doi.org/10.22323/1.270.0030
8. Lueders S (2017) Computer security: IoTs: the treasure trove of CERN. Available from: https://home.cern/cern-people/updates/2017/01/computer-security-iots-treasure-trove-cern
9. Agarwal S, Oser P, Short H, Lueders S (2017) Internet of Things security. Available from: https://doi.org/10.5281/zenodo.1035034
10. Perkel JM (2017) The Internet of Things comes to the lab. Nature 542(7639):125–126

Chapter 8
The Dark Side of Things

Licia Florio

8.1 Introduction

Over the last years the Internet of Things (IoT) has become a reality, moving from a futuristic topic to something pretty much everybody has some experience with. We are all using apps that tell us how to remain fit, and enable us to remotely control our heating, monitor health conditions and even in some cases save people's lives. These are of course only a small fraction of the endless applications of IoT.

Gartner predicts that by 2020 more than 20 billion connected "Things" will be in use [1] and some predictions are even higher. That is a lot of "Things" that will know something about us! Whilst IoT opens the doors to new opportunities (and who would not like a fridge that checks what is missing and send a reminder on what to buy?), it also creates new challenges on the security and privacy side. New Things are launched daily on the market, but how much do companies invest in securing users' access, privacy, and users' data collected by Things and often residing on the providers' cloud? Things collect an enormous amount of data that in the wrong hands may lead to dangerous data leaks.

This section provides an overview on Federated Identity Management (FIM) and its key aspects; it continues with exploring the challenges to expand the traditional federated identity management to IoT and provides an overview on how to authenticate devices and secure the way the data they collect are transmitted in a secure way.

L. Florio (✉)
Utrecht, The Netherlands
e-mail: licia.florio@geant.org

© Springer Nature Switzerland AG 2019
F. D. Hudson (ed.), *Women Securing the Future with TIPPSS for IoT*, Women in Engineering and Science, https://doi.org/10.1007/978-3-030-15705-0_8

8.2   Federated Identity Management

Identity management comprises a number of different technologies to securely enable users' access to applications. Traditionally, users would be required to create an account for each service and/or device they intended to use. Provisioning user accounts for each application that users wish to access, in combination with the increasing demand of systems that require authentication, does not scale well in a highly distributed and collaborative environment.

Over the last years, federated identity management (FIM), or federated identity and access management (FIAM), or simply federated access, has become a more pervasive way to access services. Federated identity management enables services from service providers or relying parties to consume identities that are managed by different entities known as identity providers, whilst services handle authorisation aspects. FIM basically decouples the authentication and the authorisation; in doing so it also improves the users' experience as users will not have to create accounts for each application they wish to access and consequently reduces the number of passwords they need to remember. Legal frameworks and secure technologies ensure that the identity providers and the relying parties can exchange the information in a trustworthy way and that liability aspects are covered. The main purpose of federated identity management is to allow registered users of a certain domain (such as an organisation or a campus) to safely and securely access information from other domains (such as services) in a more user-friendly way without having to provide any additional administrative user information.

This model works both in academia as well as in the commercial sector. In the research and education (R&E) sector, students get an account when they enroll at their university and can use that account to log in once and access all services that are available for that university regardless of the physical location of the service. Each institution in this case is an Identity Provider (IdP), with multiple potential Service Providers (SP) to provide services and applications to users. Institutions can also agree to share services among each other: in this case users from one institution (let us call it institution A) can access services offered by another one (let us call it institution B), by logging in at their home institution (institution A).

Identity Federations are the infrastructures deployed to enable federated access: these encompass a number of institutions that agree to interoperate and offer services under a set of well-defined rules or policies (see Fig. 8.1).

A similar approach is used also in the commercial sector. We can for instance use our Facebook or Google accounts to log into other applications (i.e. hotel reservation sites, airport Wi-Fi and many others).

Federated identity management as deployed in the R&E sector follows strict privacy laws that ensure that a very limited set of users' personal information is sent to the services.

As of May 2018, the exchange of personal data between services and identity providers in the European Union (EU) is governed by the General Data Protection

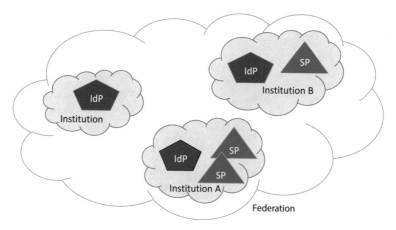

Fig. 8.1 Identity Federation Model

Regulation (also known as GDPR) [2] which has replaced the older data protection law. GDPR also regulates the transfer of personal information outside EU countries. Luckily, federated access for the R&E sector was already designed to be privacy preserving and to minimise the transfer of personal information to resource providers; the impact of GDPR is minimal on the existing federated identity management infrastructures in production in the R&E sector. The commercial sector, where data is much more sought after for profiling purposes, is quickly adjusting to GDPR to comply with it. The specific aspects of the GDPR are beyond the scope of this chapter, but it is important to note that GDPR has a big impact on IoT given the big focus on data collection and the number of third parties often involved in managing the data that are outside the EU.

8.2.1   FIM Technologies

There are different technologies to implement FIM. The choice of the technology depends on the type of resources for which the users need access. For simplicity, we list the three main approaches:

- **SAML (Security Assertion Markup Language) and OIDC (OpenID Connect)**—SAML [3] is a standard of the Organization for the Advancement of Structured Information Standards (OASIS). It is based on XML and allows the exchange of authentication information between two trusted parties, the service provider (SP) and the identity provider (IdP). OIDC [4] is the identity layer on top of the OAuth2 protocol which allows clients to verify the identity of an end-user based on the authentication performed by an Authorization Server.

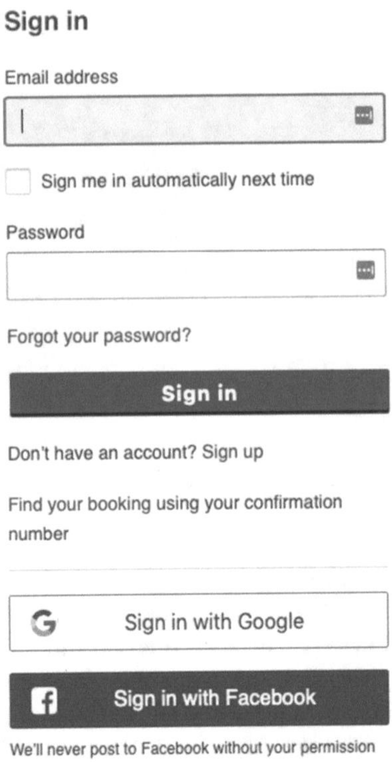

Fig. 8.2 Federated login example using Facebook or Google

Sign in

Email address

[   ]

☐ Sign me in automatically next time

Password

[   ]

Forgot your password?

Sign in

Don't have an account? Sign up

Find your booking using your confirmation number

G   Sign in with Google

f   Sign in with Facebook

We'll never post to Facebook without your permission

These two technologies enable federated access to web applications.[1] Federated access to applications is fairly common and most of us have experience with it. Think for instance about creating an account on a website that allows you to use your Facebook or Google account (Fig. 8.2) or accessing a website that allows you to use your organisational credentials (Fig. 8.3). There are differences at a technical level between SAML and OpenID Connect that concerns the type of token used to transmit authentication information and the trust model, but these differences are invisible to the user. Users normally would just use a username and password to log in to a service.

- **802.1X**—802.1X [5] is an IEEE Standard for port-based Network Access Control. It provides an authentication mechanism to devices wishing to attach to a LAN. In the academic community 802.1X is the underlying technology for eduroam, the global federated infrastructure that allows students, researchers and educators to obtain secure and free of charge network access when visiting an eduroam location. Users that can benefit from eduroam [6] can just connect their devices to the visited network without the need for temporary accounts on the guest network.

---

[1] Open ID Connect offers better support for mobile app authentication

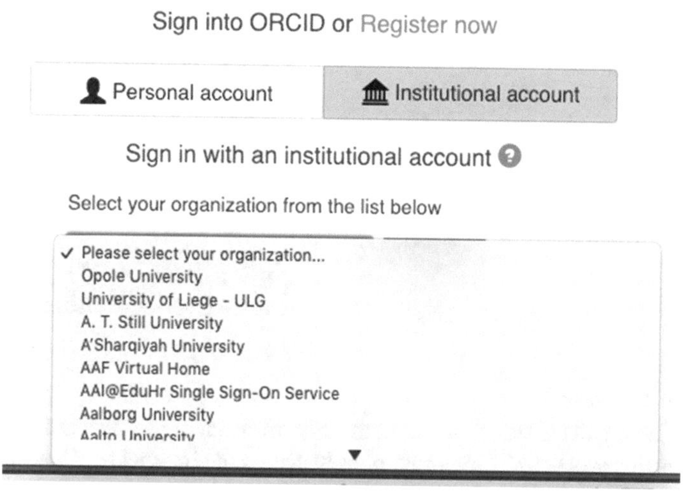

Fig. 8.3 Federated login example using organisational credentials

- **PKI-based access**—Authentication to resources can be accomplished using Public Key Infrastructure (PKI) tokens. This works by granting an application the ability to validate and verify the identity of a user based on a proof of identify which is encoded in a trusted digital certificate or equivalent PKI token. This is a fairly common mechanism used by many apps.

For completeness it is also worth mentioning cellular phone authentication. Until recently mobile phone networks and infrastructure have been quite separate from the Internet technologies. However, with phones turning more into small computers and connecting to the Internet, to other devices and servers, there is a need to bridge the gap. Most of the phones today have both cellular phone and data capability. In fact, when a phone is within range of an access point, the phone can connect to the Internet using wireless LAN (Local Area Networking). The same phone can connect also using the cellular data network and often users can choose what option to use. There is a specific protocol, EAP-SIM (Extensible Authentication Protocol— Subscriber Identity Module), defined within the IETF (Internet Engineering Task Force) that expands the EAP to leverage the SIMs in mobile phones to authenticate a client to the network.

8.3   What Is Challenging for Identity and Access Management (IAM) in IoT?

Compared to traditional IAM, in IoT the number of actors increases, including users, applications and devices. All IoT entities, that is, people, applications, services and devices, within a given ecosystem need an identity. One big difference is

Fig. 8.4 IoT Reference Model

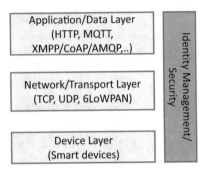

that traditional IAM is associated with the identification of physical individuals; IoT challenges this aspect as Things cannot only access the internet but they can also be accessed to gather data.

Why is it difficult to implement security and federated identity management in IoT as described above? This is due to the very nature of Things, ranging from very simple objects to more sophisticated devices; different ways to interact with them, different standards supported, and different ability to support encryption across the variety of devices.

The main threat for IoT is unauthorised users gaining access to sensitive Things and therefore data; clearly the problem is more acute if sensitive data is compromised, think for instance of healthcare or banking information. A simplified IoT reference architecture based on the ITU-T model (the Telecommunication Standardization Sector of the International Telecommunication Union (ITU)) is shown in Fig. 8.4.

All the communications that take place at different layers have to be secure and have to be able to identify a device when it interacts with other devices, applications and servers. As Fig. 8.4 shows, Identity Management and Security are orthogonal to all layers.

The main challenges for Identity and Access Management (IAM) in IoT are summarised below.

8.3.1 Device Capabilities

IoT covers a wide range of smart devices. The devices can be of different types, for instance, a wearable connected to the phone via Bluetooth, a Raspberry Pi, Google Home, home cameras, sensors, etc. The diversity of Things that users may need to access adds complexity, as it may limit the choice of technologies to manage the identity of the Things. Some IoT devices are very simple, with limited computational power, no interfaces or API to access them and can only rely on limited power (i.e. batteries). All these characteristics make it impossible to use some of the FIM protocols mentioned above, and commonly used security protocols are not always easily applicable. The IETF working group on Authentication and Authorization for Constrained Environments (ACE) [7] is working to address these aspects. The working group focuses on supporting dynamic and fine-grained access control mechanisms, where clients and/or resource servers are constrained, lack both a suitable user-interface and the ability to contact an authorization server in real-time.

The variety of devices and their distributed nature creates favourable conditions for security incidents, especially given the potential scale of the IoT network and the inherent vulnerability of some of the Things. Security is in fact the greatest concern related to IoT.

8.3.2 Device Authentication

In simple terms, a Device or Machine Authentication happens when a device (or supplicant) authenticates to the network with a stored credential. The authentication is the process where the IoT end points identify each other and verify each other's identity. Once an IoT device is on the network the device can authenticate to another one, known as machine-to-machine or M2M authentication.

It is important to note that in most cases there is no possibility to input a password into an IoT device. This means that an IoT device has to be preconfigured to be authenticated. The choice for IoT devices is rather limited; normally IoT devices can use their unique identifier, their MAC (Media Access Control) address, or, if the device allows for it, a cryptography token that can be used for authentication purposes. The identifier and/or the digital certificate are used by the device to connect to the cloud server, or to a local gateway that connects to the cloud server, and transmit information; but also, to receive information from the cloud server for instance in case of software updates.

In the case of machine-to-machine communication, authentication relies on cryptography and needs to be strong enough to protect against different types of possible attacks, such as eavesdropping, man-in-the-middle and replay attacks. The simplest form of authentication is based on a shared secret, although the distributions of the secret in a network is challenging. There is of course a risk in using this approach as if the secret is compromised the whole system is in danger. A better approach is offered by asymmetric encryption or Public Key Cryptography (PKC). This approach requires a pair of keys generated by cryptography algorithms. PKC may be challenging for small devices, as it requires available memories and sufficient computational power. Beside the traditional public key algorithms, elliptic-curve cryptography (ECC) [8] is a most promising technology for IoT as it uses smaller key sizes and requires less computational power, characteristics that better suit the capabilities of IoT devices.

Conversely, user authentication always involves a human that interacts with a system and prompts for either username and password or some other type of tokens. It is possible to deploy more sophisticated approaches than human interaction to improve security, such as period password regeneration and digital certificates renewals. However, these approaches are challenging in an IoT network with a high level of devices, as a new certificate or a cryptography would have to be provisioned into a device.

8.3.3 Device Connection to the Internet

The confidentiality and security of the data is one of the major challenges in the IoT, particularly in the consumer market, where users rely on many different smart devices. The volume of data collected by IoT devices and stored in the cloud can contain sensitive information which, in some cases, are sent to unknown servers. To this end it is important to distinguish between devices that do not directly connect to the internet, such as wearable devices and simple sensors, and more sophisticated devices capable of connecting to the Internet directly, such as smart car systems, smart home devices such as Google Home and Amazon Echo.

In the case of a wearable device, the device generally connects to an app on the user's mobile phone to transmit data. The communication between the device and the app happens generally via Bluetooth, which is a fairly secure protocol. The app in turn connects to a cloud server to store the user's data. It is important the communication between the app and the server happens in a secure way, to prevent data sniffing, that is, an unauthorised party that captures and reads the data transmitted. Other IoT devices may connect to a gateway, that is, a bridge between IoT devices and the cloud server. A gateway can also collect information from several IoT devices, aggregate it and send it to the cloud server. A gateway provides some benefits; it can reduce the "transmission burden" on the IoT device boosting its battery life; it can translate the data collected from different sensors into a standard data protocol; it can also improve security by effectively preventing the need for devices to be on the Internet (Fig. 8.5).

The communication between IoT gateways and the cloud server happens via different protocols; the most well-known are MQTT (MQ Telemetry Transport or Message Queuing Telemetry Transport), CoAP (Constrained Application Protocol), XMPP (Extensible Messaging and Presence Protocol), ADMQP (Advanced

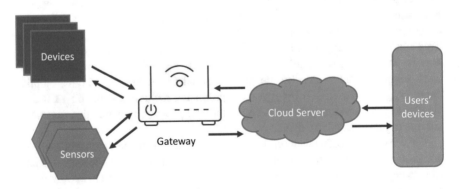

Fig. 8.5 IoT gateways

Message Queuing Protocol) and HTTP all of which have different characteristics and security features. It is important to note that the size and the lower power processor of many IoT devices pose challenges regarding the type of protocol the device can support and consequently the encryption the device is capable of; the latter is important to ensure that data is securely transmitted.

8.3.4 Scale

The number of devices connected to a gateway or controller has an impact on the authentication demand in IoT where multiple devices need to authenticate to a gateway and the gateway needs to authenticate to a server. Traditional methods may not scale well enough to keep up with the demand or may result in excessively expensive solutions. More devices connected means more data, more connections to send the data and more authentication required. The scale of the devices connected poses security challenges. With many devices connected, a system is as secure as its weakest link; and thanks to the many devices connected, security incidents and DDoS (Distributed Denial of Service) attacks can propagate much faster than in a traditional scenario. A quick search on the Internet reveals a variety of security incidents, ranging from smart meters used by utility companies to monitor the home consumption, to car systems, to home systems, all being hacked via device vulnerabilities. Also, big companies like Google and Amazon have not been immune: a researcher demonstrated how to take over Amazon Echo using vulnerabilities in Amazon Echo devices [9]. If this is a concern for every day appliances, such an attack becomes a serious threat for medical devices. There are some measures that can mitigate risk, like selecting devices that offer support for better encryption, deploying well-known protocols and phasing out legacy technologies when designing an IoT network.

8.4 Authentication of Users in IoT

In the consumer sector, users generally interact with a limited number of IoT devices at the same time. The users' interaction with the IoT devices is not always active and, in many cases, does not even require authentication.

Typically, the interaction happens via an app on a smart phone or via specific devices, such as a thermostat console to monitor the house heating. Users create an account and register their account with a specific device. However, with more devices connected, users need to create and juggle multiple accounts, which makes the process inconvenient for the users.

For other scenarios it is desirable to have a more direct interaction for the user to control who can access and process their data. The technology is maturing to make such an interaction not only possible but also secure, via biometric mechanisms and multifactor authentication (MFA).

With more devices connected, some big companies have joined forces to create an environment to make it possible to aggregate information across various connected devices. An example of this is Apple Home Kit that allows users to control different appliances via an Apple iOS device. Although this is still far from the larger scale Single Sign On (SSO) that would be desirable, and does not remove silos across vendors, it is certainly a step in the right direction.

8.4.1 Access to IoT Medical Devices

In the previous chapters, we have mentioned that the scale of IoT devices connected to networks and each other poses security challenges.

The talk of Dr. Marie Moe at TNC2018 [10] was particularly enlightening with regard to security and authentication to IoT devices used for medical purposes. Marie Moe suffered from a heart condition, as a result she needed a pacemaker. However, as a trained security researcher she started to investigate the security vulnerability in the software used by the pacemaker as well as the possibility of hacking the device. Her specific pacemaker could be connected to the network via a WiFi connection and monitored remotely. Via the remote access, patients' data are retrieved daily and sent to the vendor, and via that to the healthcare professional for statistics and monitoring purposes. Whilst this is a great feature to monitor the health of a patient, it raises questions on how the data are protected. The most interesting aspect was that access to the pacemaker data was not protected, therefore in theory everybody could access these sensitive data. Dr. Moe also reported that pacemakers can be hacked and that they are vulnerable to man-in-the-middle attacks. The regulations in this area are still fairly new; there are clearly many regulations for medical devices; however, security considerations are only just being addressed.

8.5 Which Approach Works Better for IAM in IoT?

This chapter describes the current approaches to FIM and the challenges in deploying FIM in IoT. We have presented different standards and frameworks, that can be deployed, but all of them have their limitations when used in an IoT ecosystem.

SAML is a robust protocol with a big footprint in academia and in large companies; it addresses many aspects needed to deploy IAM, including identity federation and SSO functionality. However, its complexity and lack of support in some of the IoT devices limit its usefulness for IoT.

OIDC and the underlying OAuth2 offer better support for mobile device authentication and could also support more direct privacy control on the users' side. OIDC also provides built-in discovery and registration mechanisms that are useful in scaling any architecture to the numbers of actors that IoT can create.

SIM and mobile phones (3G and 4G) offer a good authentication approach, also allowing implementation of multi factor authentication, given the ubiquity of the phone and the fact that users always carry them. However, not all IoT devices connect to phones.

802.1X can be used to on-board IoT devices in a network, assuming that a device is capable of supporting the protocol. This could be very useful for instance on a university campus that most likely already deploys an 802.1X network and may need to add IoT devices for instance to monitor parking spots, temperature, lights, etc.

Using PKI tokens, whether in the form of digital certificates or tokens, provides a strong authentication system. However, in the IoT, many devices may not have enough memory to store a certificate or may not even have enough required CPU power to support cryptographic functions to validate the X.509 certificates.

There is no simple answer, nor one answer, regarding the best approach for Identity and Access Management (IAM) for IoT. Each choice has to be evaluated in the context of the needs that an IoT network has to support and the type of devices to connect.

In summary, authentication and authorisation of IoT actors is essential to IoT security. Standards do exist to allow for authentication and authorisation at different communication layers and also to allow the user to consent to sharing their data, although they are not built in from the start. Regulation in IoT is slowly happening and users' awareness on privacy in the IoT space is also increasing; these factors will apply pressure on the IoT industry to reduce the risk in the dark side of things.

References

1. https://www.informationweek.com/mobile/mobile-devices/gartner-21-billion-iot-devices-to-invade-by-2020/d/d-id/1323081
2. https://en.wikipedia.org/wiki/General_Data_Protection_Regulation
3. https://en.wikipedia.org/wiki/Security_Assertion_Markup_Language
4. https://openid.net/connect/
5. https://en.wikipedia.org/wiki/IEEE_802.1X
6. https://www.eduroam.org/
7. https://datatracker.ietf.org/doc/draft-ietf-ace-actors/?include_text=1
8. https://ieeexplore.ieee.org/document/7993462
9. https://www.wired.com/story/hackers-turn-amazon-echo-into-spy-bug/
10. https://tnc18.geant.org/web/media/archive/1A

Chapter 9
Public Safety and Protection by Design: Opportunities and Challenges for IoT and Data Science

Alicia D. Johnson, Meredith M. Lee, and Soody Tronson

9.1 Introduction

The rapid growth and integration of technology in our lives present new challenges and opportunities for our communities. In particular, the Internet of Things (IoT), with a connected network of sensors and computing power, links disparate elements for new functionality and coordination. From connected vehicles that could streamline transportation in smart cities [1–3], to early warning systems for earthquakes or global health hazards [4–6], IoT and data-enabled technologies have the potential to impact public safety at an unprecedented scale.

Yet, our legal structures, operating frameworks, and social norms concerning the design and deployment of IoT—and the collection, use, and sharing of data—often lack the transparency and cohesiveness needed for effective coordination. Especially for matters of public safety and social good, we need to collectively understand and proactively shape an ecosystem that leverages the democratization of technology and values collaboration among diverse stakeholders.

Although research is underway in privacy-preserving analytics [7–9], the complex social context in which such approaches would be deployed requires a new level of multi-stakeholder collaboration. Analytics and decision making processes often need to be collaborative, providing transparency and accountability, while

A. D. Johnson
Sonoma, CA, USA
e-mail: adj032008@gmail.com

M. M. Lee (✉)
Berkeley, CA, USA
e-mail: mmlee@berkeley.edu

S. Tronson
Woodside, CA, USA
e-mail: soody@stlgip.com

© Springer Nature Switzerland AG 2019
F. D. Hudson (ed.), *Women Securing the Future with TIPPSS for IoT*, Women in Engineering and Science, https://doi.org/10.1007/978-3-030-15705-0_9

assessing the issues of inherent bias and the potentially competing goals among entities. In IoT and data science, information and power are often distributed by design to more than one entity. In a similar vein, whether through technology pilot programs or public-private partnerships, governments and other organizations securing the safety of the public are increasingly interfacing across disciplines to be responsive to operational needs. By drawing insight from computer science, law, government operations, and other areas, we have an opportunity to highlight a shared responsibility, and to shape the responsible and ethical implementation [10–13] of public safety solutions and approaches.

In this chapter, with safety defined as the "protection from risk or injury," we note the broad interpretations of the word *Protection*. Protection can encompass the safeguarding of human lives, property, or infrastructure, as well as legal and other formal measures intended to preserve civil liberties and rights. With new developments constantly emerging in IoT and data science, one must consider not only the sustainability of cyber-physical security and safety protections, but also more nuanced issues surrounding privacy, intellectual property, and intended use. The lack of cohesion from existing federal, state, or other laws and guidance poses a significant challenge for navigating the future of IoT with the backdrop of technological and social change.

Like the other TIPPSS topics of Trust, Identity, Privacy, and Security, Safety and Protection are global, collective challenges in today's interconnected society, with opportunities for local action. Key to the sustainability of solutions and approaches will be the insights about stakeholder risks and contexts, and the creation of avenues for clarifying and improving policies and processes.

9.2   The Evolving Landscape of Safety

IoT represents some of the most progressive and deliberate integration of technology into the field of public safety and disaster science in recent memory. From early earthquake warning technologies [14] to the exploitation of this technology by former domestic partners [15], IoT has found itself integrally linked to safety and security.

The use of alert and warning technologies with low latency designed to benefit communities provides state-of-the-art opportunities to communicate efficiently and effectively with populations in need. The Wireless Sensor Networks (WSN) built to support daily IoT integration continue to develop. Metropolitan areas are now seeing tremendous growth in network stability, which leads to more rapid alert mechanisms. Rural areas have yet to receive robust network development, thereby setting their populations at a disadvantage when mass notifications are necessary. While governments often note IoT's one-way notification capabilities, it is clear that communication between residents, tourists, and responders in a disaster impacted area is possible and highly likely as IoT capabilities increase.

In addition to the gap between cities and their rural counterparts, one must also consider that ubiquitous IoT has a large socioeconomic hurdle to overcome. Individuals and communities unable to place or purchase IoT-enabled devices in their homes, vehicles, community buildings, or on their person may likely receive slower alert and notification in a crisis or disaster. This lack of ubiquity is troubling when it comes to protecting people and places we value. Moreover, surveillance and privacy concerns are far from neutral. Whether instigated by the government or private sector, surveillance is often colored by ethnicity and gender. After 9/11, the US Government worked closely with the New York City Police Department to surveil ethnic communities, while a century earlier, police heavily monitored leaders of the women's suffrage movement, all in the name of public safety and security [16]. This type of monitoring, particularly using IoT products that are often owned by the user themselves, breeds fear and distrust by populations most likely to be targeted. In the end, that distrust hampers the saturation needed to create a network worthy of rapid, immediate disaster notification.

The creation of smartphone applications [17] as well as technologies designed to automate actions when a set of criteria is met, are becoming a priority for cities throughout the world. Whether a community has an application dedicated to disaster alerts or relies on IoT to automatically stop elevators in anticipation of an earthquake, IoT is everywhere and often operating unnoticed in the background of everyday systems. Given the small number of disaster scenarios able to accurately test such mechanisms, reliability and ability to motivate the public towards appropriate action could prove challenging.

The use of IoT in public safety can be far more intimate than large scale alert and notification. In domestic violence cases, courts are experiencing a rise of victim targeting by way of connected smart devices. IoT products such as locks, speakers, thermostats and cameras are convenient for home management, but are also being used by domestic partners as a mechanism for harassment, unwanted monitoring, and control. These devices often provide a window into a location where privacy would be expected, and are increasingly being used to torment and victimize [15]. As with large-scale surveillance concerns, this type of use case is often skewed by gender, ethnicity, and socioeconomic background [18].

The abuse of IoT may never be fully remediated, but with careful design and feedback from impacted stakeholders, the technologies could enable more effective information-sharing—in times of crises as well as in routine operation. Trust must be established between the technology owner, the technology user, and any third party seeking to employ the technology, in order to create a network robust enough to alert populations of impending risk of injury. Those seeking to deploy IoT technologies widely for alert and notification in large-scale disasters acknowledge a critical mass of users that trust the capability and security of the technologies must be reached to provide sufficient penetration for actionable notification. Governments and technology developers may consider implementing "opt-out" opportunities, similar to the European Union's General Data Protection Regulation (GDPR) "right to be forgotten" that could accelerate adoption and offer opportunities for privacy. Additionally, technology developers may consider how data use and retention impact the ability to respond to and prepare for any public safety concern. As revelations about data breaches and misuse continue to surface, our communities have an opportunity for conversation and action to strengthen shared understanding, ask questions, and help shape the trajectory of how IoT and data science impact our lives.

9.3   Law and Practice: Need for Coordination in a Complex Ecosystem

The General Data Protection Regulation (GDPR) effective May 25, 2018 [19], designed to "harmonise" data privacy laws across Europe, reconstructed how businesses process and handle data. In the USA, however, there is currently no single principal data protection legislation or law relating to protection and rights of individuals, or regulating the collection and use of personal data.

Presently in the USA, there are approximately 20 national-scale privacy or data security laws [20–29] and hundreds of locally applicable laws among the 50 states and territories [30–32]. The federal laws are largely specific to certain sectors such as finance and health, while the existing state laws focus on privacy rights of individuals, generally requiring notice for pre-collection and opt-out and opt-in for use of regulated personal information. While the right to personal privacy has been present under common law, statutes in different states vary regarding what information is protectable. Certain federal laws preempt state laws relating to the same topic [20], while others do not. Article VI, Clause 2 of the US Constitution (The Supremacy Clause) [33] contains what is known as the doctrine of preemption, which specifies that the federal government wins in the case of conflicting legislation. However, a lack of consistency between federal and state laws, and between states, can be seen in cases spanning consumer privacy [30], net neutrality [34], health insurance, immigration [35], employment [35], and more. Both consumers and businesses can be impacted by this lack of regulatory certainty as they move or transact between states. It is not difficult to imagine how issues can arise from the lack of coordination and consistency as IoT is deployed in an increasingly global and connected community.

Among the USA, at least 24 states currently have laws that address data security *practices* of private sector entities. Most of these data security laws require businesses that own, license, or maintain personal information about a resident of that state to implement and maintain "reasonable security procedures and practices" appropriate to the nature of the information and to protect the personal information from unauthorized access, destruction, use, modification, or disclosure [31]. All states have security measures in place to protect data and systems, with at least 19 states at present requiring, by statute, that state government agencies have in place specific policies or measures to ensure the security of the data they hold. At least 35 states and Puerto Rico have enacted laws that require either private or governmental entities, or both, to destroy, dispose of, or otherwise make personal information unreadable or undecipherable [32]. Even among those states that have data disposal laws, whether the laws apply to government or just to businesses varies.

By way of example, California currently has more than 25 state privacy and data security laws [36]; while not as expansive as Europe's GDPR, these laws lead the nation in restricting how technology companies collect, store, and use personal data. The California laws in this area include the New Electronic Communications Privacy Act (CalECPA) [37], and the California Consumer Privacy Act of 2018 (CCPA) (effective 2020) [38]. CalECPA was enacted in response to increased concerns about government access to consumer digital information and the exploitation of privacy laws to turn mobile phones into tracking devices and points of access to emails, digital documents, and text messages without proper judicial oversight. CalECPA prohibits a government entity from compelling the production of or access to electronic communication information or electronic device information, as defined in statute, without a search warrant, wiretap order, order for electronic reader records, or subpoena issued pursuant under specified conditions, except for emergency situations. As defined in the statute, an emergency involves danger of death or serious physical injury to any person. As with many laws, there are exceptions to exceptions. CalECPA specifically excludes several methods of obtaining electronic information from its warrant requirement [39].

The CCPA was introduced in the California legislature to address provisions set forth in a ballot initiative that would have enacted a different and distinct California Consumer Privacy Act. It was passed on June 28, 2018 and will be effective on January 1, 2020 [38]. The CCPA, as passed, creates four basic rights for California consumers: (1) A right to know what personal information a business has about them, and where (by category) that personal information came from or was sent; (2) A right to delete personal information that a business collected from them; (3) A right to opt-out of the sale of personal information about them; (4) A right to receive equal service and pricing from a business, even if the consumer exercises their privacy rights under the Act, but with significant exceptions. While the right-to-know extends to all information a business has collected about a consumer, the right-to-delete only applies to the information a business collected from the consumer. According to analysis by Electronic Frontier Foundation (EFF), CCPA, as passed, however, suffers from several deficiencies, notably: (1) The Act allows businesses to charge a higher price to users who exercise their privacy rights; (2) The Act does not provide users the power to bring violators to court, with the exception of a narrow set of businesses if there are data breaches; (3) For data collection, the Act does not require user consent; and (4) For data sale, while the Act does require user consent, adults have only opt-out rights, and not more-protective opt-in rights; (5) The Act's right-to-know should be more granular, extending not just to general categories of sources and recipients of personal data, but also to the specific sources and recipients [40].

These new laws still fall short of the GDPR protections. The scope of disclosures required by the GDPR extends beyond that required by the CCPA, and CCPA's deletion right applies only to data collected from the consumer (i.e., not to data about the consumer collected from third party sources), whereas the GDPR's applies to all data concerning a data subject. Consequently, entities that are subject to both laws (California and GDPR) may see certain differences between the qualifying data and its treatment under the two regimes. This will lead the entities to establish different mechanisms for compliance with each of the different regimes or a mechanism that conforms to the most common rigorous denominator.

In reviewing these differences in various laws, the definitions of certain terms also vary greatly from law to law within the USA. For example, the definition of personal data varies greatly by regulation and state. The Federal Trade Commission (FTC) considers information that can reasonably be used to contact or distinguish a person, such as Internet Protocol (IP) addresses and device identifiers, as personal data. On the other hand, very few US federal or state privacy laws define "personal information" as including information that on its own does not actually identify a person. The definition of sensitive personal data also varies widely by sector and by type of statute. Generally, data pertaining to personal health, student records, personal information collected online from children under 13, finances including credit worthiness, and information that can be used to carry out identity theft or fraud are considered sensitive. These variations in legal definitions create additional challenges as IoT and data technologies begin to traverse sectors.

Furthermore, to the extent that laws regarding security and privacy of personal data exist, they vary, and most do not address many of the issues surrounding the gathering, maintaining, and use of personal data or information. As such, these issues are left to private entities that enter into relationships involving data and its processing. For example, in certain states, a business engaging another business to process gathered personal data must bind the processor to take reasonable measures to protect the security of the personal data. The form and particularities of such an agreement in these laws, however, are not specified. This lack of cohesiveness opens the door for varied interpretations and litigation. Therefore, a given entity may face a complex array of data categories and the various applicable federal and state laws. Many third party services have emerged to respond to this complexity, but not necessarily with uniformity—highlighting an opportunity for future efforts focused on trustworthy, more universally applicable standards and certification schemes.

9.4   Data as Knowledge-Based Assets

The increasing prominence of data has been highlighted as part of the "Fourth Industrial Revolution" [41]. Data are assets and can be created, manufactured, processed, stored, transferred, licensed, sold, and stolen. In 2017, the legal buying, selling, and trading of personal medical data totaled $14 billion USD [42]. With the context of various legal and ethical requirements regarding the treatment of personal data, one of the pressing questions for many stakeholders is how data can be shared via contractual relationships.

Data can be seen as an *intangible* asset, and like other knowledge-based assets, they can be reproduced and transferred at low marginal cost. However, as opposed to the concept of ownership of physical goods, where the owner typically has exclusive rights and control over the goods (including, for instance, the freedom to destroy the goods), this is not the case for intangibles such as data.

For intangible assets, intellectual property rights (IPRs) are typically suggested as the legal means to establish clear ownership. There is, however, a main distinction between traditional forms of IPRs (e.g., copyrights, patents) and those forms involving data. In the case of the traditional IPRs, statutes confer exclusive rights to exploit that right. For example, copyright laws give the owner the exclusive right to exploit that copyright (excluding limitations such as fair use), and in the case of patents, only the owner can exploit the property which is the subject of the patent.

Thus, all other entities who wish to exploit the asset must obtain an affirmative specific right or license to exploit the IP rights held by the owner. Exploitation of anything else which has not been specifically and affirmatively authorized, remains prohibited by law. However, in the case of data, legal regimes such as copyright as well as other IPRs applicable to databases and trade secrets can be used only to a limited extent [43].

Although technologies such as cryptography have dramatically reduced the costs of exclusion and thus are often used as a means to protect data, there are additional challenges with respect to different stakeholders having different rights. For example, some stakeholders have "the ability to access, create, modify, package, derive benefit from, sell or remove data, but also the right to assign these access privileges to others." In cases where the data are considered "personal data," the situation is even more complex, since certain rights of the data subject cannot be waived [43].

To better understand how to navigate the legal ecosystem, several issues need to be resolved, including:

- *What are the type of data and their classification?*
- *Broadly, what rights attach to data? When and under what circumstances do these rights attach?*
- *Who has control of the data based on these constraints?*
- *Who owns the data? What rights can the owner exercise over the data* [44]*?*
- *Are there any property rights in the raw data or knowledge derived from the data? Are there any rights associated with the selection, coordination, and arrangement of the data?*

Typically and historically, data generated by businesses in the private sector (e.g., a company's sales information) are owned by the business entity. The breadth of personal information that can now be easily generated and captured has raised issues about ownership and the lack of protective laws to expressly define ownership. For example, who owns an individual's medical data? Many patient consent forms for the performance of tests or operations state that all the data or tissue samples belong to the doctor or institution performing the procedure [45] and courts have generally ruled against patients in disputes over ownership of human tissue [46]. Even if steps are taken to protect identities, data can be the subject of transactions without any notification to the patient [47].

The scope of various regulations regarding data ownership and portability may be limited to raw personal data provided by the data subjects, but may also include *results* that are deduced or derived from the raw data. The transformation of data presents additional challenges as there is a lack of uniformity in laws that comprehensively enumerate the rights of the owner of the original data and the rights of the data aggregator or analyst—creating a system vulnerable to confusion and misuse.

For example, while "raw data," to the extent purely factual, should not be subject to any copyright protection [48], its compilation in a database via selection, coordination, and/or arrangement of data, information, or files may be subject to copyright protection [49]. Similarly, another body of law, that of trade secrets, may also pertain to ownership issues about any raw data originally created by a business, as well as any *results* from such raw or processed data. In the USA, trade secret law has been primarily in the purview of individual states, with each state having substantial variations including some significant variation on the definition of a trade secret. To address lack of uniformity among states, the Uniform Law Commission published the Uniform Trade Secrets Act (UTSA) in 1979 and amended the Act in 1985. The UTSA defines trade secrets and claims related to trade secrets; the Act was promulgated for adoption in the USA.

Yet, while an overwhelming majority of states have enacted variations of the UTSA, there has been a degree of uncertainty as a result of legislative action and state court interpretations of UTSA laws. For example, state law exhibits significant variation in the definition of a trade secret. Most recently, the Defend Trade Secrets Act (DTSA) of 2016 introduced a new federal civil cause of action for trade secret misappropriation. DTSA also adopted the definition of a trade secret to expressly include source code, algorithms, programs, and data sets as potential types of information that may be considered a trade secret.

As the laws, frameworks, and case studies involving IoT and data continue to develop, it is imperative that community stakeholders discuss the local and global impacts of new policies.

9.5 Questions and Opportunities for Action

IoT and data science are beginning to reshape public safety, through a combination of deliberate as well as "ad hoc" efforts. As with any new technological process, definitions are critical to operations and engagement with these technologies.

When seeking to leverage IoT and data, one can consider question-focused frameworks to guide stakeholder discussions beyond the topic of compliance issues. Particularly for topics relating to public safety and protection, clearly identifying and communicating salient details to the public in a timely manner can help build the trust needed in these applications. Concepts of responsible and ethical actions are key throughout this framework:

- *What is the purpose and intended use?*
- *Have we clearly described the type and scope of any data, devices, and systems?*
- *What are the rules of engagement relevant to the data, devices, and systems?*
- *What roles and responsibilities exist for each stakeholder, related to processing, maintaining, and sharing any data?*
- *Under what conditions can entities publish and disseminate results?*
- *How might we proactively mitigate the risk of unauthorized access or unwarranted use?*

Increasingly, in the public and private sectors, fields ranging from the physical and social sciences to engineering and business are becoming data-intensive; both commercial success and academic impact are often dependent on having access to data. Many organizations collecting data lack the expertise required to process it, and, thus, pursue data sharing with researchers who can extract more value from the data. At the same time, researchers often search for real-world data sets to understand and improve the effectiveness of new methods and innovations. Unfortunately, many data sharing attempts fail, for reasons ranging from legal restrictions on how data can be used, to privacy policies, different cultural norms, technological barriers, cumbersome processes in reaching agreement, and differing negotiation power. Data sharing partnerships that are vital to addressing pressing societal challenges in cities, health, energy, and the environment are not being pursued due to such obstacles.

Pragmatically addressing our community's data sharing challenges requires open, supportive dialogue across many lines of effort, including technology, policy, research, and operations. Furthermore, there is a crucial need for well-defined agreements that can be shared among key stakeholders, including researchers, technologists, legal representatives, and technology transfer officers.

There are already standards, such as ISO 27001, to help to achieve GDPR compliance, paving the way for increased discussion about initiatives, standards, and guidance on IoT and data-related technologies. To streamline data sharing practices, such standards and initiatives must provide a flexible framework that can accommodate the needs of particular transactions and the evolution of future laws, regulations, and court opinions. A goal, in developing such resources, should be to preserve and enhance efficient and ethical practices that promote the generation of value across a broad range of stakeholders—whether they be consumers or suppliers of data. When developing requirements and guidance to facilitate ethical conduct, we have an opportunity to not only incorporate quantitative and qualitative metrics, but also to purposefully highlight the nuanced reflection and dialogue resulting from experience.

The relationship between public safety, IoT, and data science must be built on trust. To facilitate sustainable approaches and solutions, mechanisms for data collection, analysis, and sharing must recognize the complex ethical and human-centric implications of work in this arena. Educational resources and training should focus on the "translational" capabilities and impacts of data science, emphasizing real-world applications within a broad societal ecosystem, pathways for feedback, and team-based contributions.

References

1. National Academies of Sciences, Engineering, and Medicine (2018) Critical issues in transportation. The National Academies Press, Washington, DC, p 2019. https://doi.org/10.17226/25314
2. Lu N, Cheng N, Zhang N, Shen X, Mark JW (2014) Connected vehicles: Solutions and challenges. IEEE Internet Things J 1(4):289–299
3. Mulligan DK, Bamberger KA (2016) Public values, private infrastructure and the Internet of Things: the case of automobiles. Journal of Law & Economic Regulation, Vol. 9. No. 1. 2016. 5. pp.7–44, https://escholarship.org/uc/item/3z59j68j

4. Ray PP, Mukherjee M, Shu L (2017) Internet of things for disaster management: state-of-the-art and prospects. IEEE Access 5:18818–18835
5. Islam SR, Kwak D, Kabir MH, Hossain M, Kwak KS (2015) The internet of things for health care: a comprehensive survey. IEEE Access 3:678–708
6. Chin CD, Linder V, Sia SK (2007) Lab-on-a-chip devices for global health: Past studies and future opportunities. Lab Chip 7(1):41–57
7. Dwork C, Roth A (2014) The algorithmic foundations of differential privacy. Found Trends Theor Comput Sci 9(3–4):211–407
8. Sweeney L (2002) k-anonymity: a model for protecting privacy. Int J Uncertain Fuzziness Knowl Based Syst 10(05):557–570
9. Popa RA, Balakrishnan H, Blumberg AJ (2009) VPriv: Protecting privacy in location-based vehicular services, https://dspace.mit.edu/openaccess-disseminate/1721.1/58903Cached
10. Erickson L, Evans Harris N, Lee MM (2018) It's time for data ethics conversations at your dinner table. Tech at Bloomberg
11. Gotterbarn DW, Bruckman A, Flick C, Miller K, Wolf MJ (2018) ACM code of ethics: a guide for positive action. Communications of the ACM 61(1):121–128 10.1145/3173016, https://cacm.acm.org/magazines/2018/1/223896-acm-code-of-ethics/abstract
12. Angwin J, Larson J, Mattu S, Kirchner L (2016) Machine bias. ProPublica, https://www.propublica.org/article/machine-bias-risk-assessments-in-criminal-sentencing
13. Stoyanovich J, Howe B, Abiteboul S, Miklau G, Sahuguet A, Weikum G Fides: Towards a platform for responsible data science. Proceedings of the 29th International Conference on Scientific and Statistical Database Management; ACM, 27 June 2017, p. 26
14. Khalil IM, Khreishah A, Ahmed F, Shuaib K (2014) Dependable wireless sensor networks for reliable and secure humanitarian relief applications. Ad Hoc Netw 13:94–106
15. Bowles N (2018) Thermostats, locks and lights: digital tools of domestic abuse. The New York Times
16. Sherman J (2018) Need a resolution? How about 'Guard your online presence'. Richmond Times-Dispatch
17. Lin R (2019) Long-awaited earthquake early warning app for L.A. can now be downloaded. Los Angeles Times
18. Rosen D, Singh S (2018) The New America Foundation. Perspectives and Policies on the Digital Safety of Vulnerable Communities
19. General Data Protection Regulation. https://ec.europa.eu/commission/priorities/justice-and-fundamental-rights/data-protection/2018-reform-eu-data-protection-rules_en
20. The Financial Services Modernization Act (Gramm-Leach-Bliley Act (GLB)) (15 U.S.C. §§6801-6827)
21. The Health Insurance Portability and Accountability Act (HIPAA) (42 U.S.C. §1301 et seq.)
22. The Federal Trade Commission Act (FTC ACT) (15 U.S.C. §§41-58)
23. The Driver's Privacy Protection Act of 1994 (18 U.S. Code 2721)
24. The Fair Credit Reporting Act, as amended by Fair and Accurate Credit Transactions Act (FACTA) (15 U.S. Code 1681)
25. The Controlling the Assault of Non-Solicited Pornography and Marketing Act (CAN-SPAM Act) (15 U.S.C. §§7701-7713 and 18 U.S.C. §1037)
26. Telephone Consumer Protection Act (47 U.S.C. §227 et seq.)
27. Children's Online Privacy Protection Act (COPPA) (15 U.S. Code 6501)
28. Video Privacy Protection Act (VPPA) (18 U.S. Code 2710)
29. The Electronic Communications Privacy Act (18 U.S.C. §2510) and the Computer Fraud and Abuse Act (18 U.S.C. §1030)
30. National Conference of State Legislatures (2019) State Laws Related to Internet Privacy. http://www.ncsl.org/research/telecommunications-and-information-technology/state-laws-related-to-internet-privacy.aspx
31. National Conference of State Legislatures (2019) State Laws Related to Data Security Laws in Private Sector. http://www.ncsl.org/research/telecommunications-and-information-technology/data-security-laws.aspx

32. National Conference of State Legislatures (2019) State Laws Related to Data Disposal. http://www.ncsl.org/research/telecommunications-and-information-technology/data-disposal-laws.aspx
33. United States Senate. Constitution of the United States. https://www.senate.gov/civics/constitution_item/constitution.htm#a6
34. United States of America v. State of California et al. 2:18-cv-02660 (US District Court for the Eastern District of California. Oct 3, 2018)
35. United States of America v. State of California, et al. 2:18-cv-00490-JAM-KJN (US District Court for the Eastern District of California. July 9, 2018)
36. State of California Department of Justice. Privacy Legislation Enacted in 2013. https://oag.ca.gov/privacy/privacy-legislation/leg2013
37. The State of California. The California Electronic Communications Privacy Act. https://leginfo.legislature.ca.gov/faces/billNavClient.xhtml?bill_id=201520160SB178
38. The State of California. The California Consumer Privacy Act of 2018. https://leginfo.legislature.ca.gov/faces/billTextClient.xhtml?bill_id=201720180AB375
39. State of California. Privacy: electronic communications: search warrant. http://leginfo.legislature.ca.gov/faces/billNavClient.xhtml?bill_id=201520160SB178
40. How to Improve the California Consumer Privacy Act of 2018. Electronic Frontier Foundation. August 8, 2018. https://www.eff.org/deeplinks/2018/08/how-improve-california-consumer-privacy-act-2018
41. World Economic Forum. The Fourth Industrial Revolution: what it means, how to respond. January 14, 2016. Available at https://www.weforum.org/agenda/2016/01/the-fourth-industrial-revolution-what-it-means-and-how-to-respond/
42. Global Big Data in Healthcare Market: Analysis and Forecast, 2017-2025 (Focus on Components and Services, Applications, Competitive Landscape and Country Analysis). BIS Research. March 2019. https://bisresearch.com/industry-report/global-big-data-in-healthcare-market-2025.html
43. OECD, KEY ISSUES FOR DIGITAL TRANSFORMATION IN THE G20, 150–62 (2017). Available at http://www.oecd.org/innovation/key-issues-for-digital-transformation-in-the-g20.pdf
44. Ritter J, Mayer A (2017) Regulating data as property: a new construct for moving forward. Duke Law Tech Rev 16:220
45. Petrow S. Who owns your medical data? Most likely not you. The Washington Post. 2018 November 25
46. Dry S (2009) Who owns diagnostic tissue blocks? Lab Med 40(2):69–73
47. Moore v. Regents of the University of California, 51 Cal. 3d 120; 271 Cal. Rptr. 146; 793 P.2d 479
48. Feist Publications, Inc. v. Rural Telephone Service Co., Inc., 499 U.S. 340, 350, (1991)
49. Copyright Act, 17 U.S.C. 101. Available at https://www.govinfo.gov/app/details/USCODE-2011-title17/USCODE-2011-title17-chap1-sec101

Chapter 10
Privacy Management in the Internet of Things (IoT)

Grace Wilson Caudill

10.1 Introduction

Data Security and Privacy concerns are the two big concerns in the Internet of Things. The Internet of Things (IoT) can be likened to a "child node" of a "parent node," in that IoT is an extension of the Internet. IoT is an integration of mobile networks, the Internet, social networks, and intelligent devices to provide better services or applications to users. The IoT connect the digital cyberspace and real physical space, in which radio-connected intelligent sensors have invaded the physical space, and these are now embedded in everything from our toys, to our office equipment, to our healthcare medical devices. It is quite evident that the IoT can introduce all the vulnerabilities of the digital world into our real world. Privacy risks will arise as objects within the IoT collect and aggregate fragments of data that relate to their service(s). The collation of multiple points of data can swiftly become personal information as events are reviewed in the context of location, time, recurrence, etc. This is one aspect of the big data challenge, and security professionals will need to ensure that they think through the potential privacy risks associated with the entire dataset of big data.

New technologies bring new benefits, conveniences and improvements in the quality of life; however, with these advantages come disadvantages and challenges. The Internet of Things (IoT) can produce massive amounts of data. This data must be transmitted, processed in some way, and then potentially stored somewhere, hopefully securely. Much of this data is personal data, or can be combined to become Personal Identifiable Information (PII), and some can be quite sensitive. Personal Identifiable Information can be any representation of information that permits the identity of an individual to whom the information applies to be reasonably inferred;

G. Wilson Caudill (✉)
Lexington, KY, USA
e-mail: nosliw357@yahoo.com

© Springer Nature Switzerland AG 2019
F. D. Hudson (ed.), *Women Securing the Future with TIPPSS for IoT*, Women in Engineering and Science, https://doi.org/10.1007/978-3-030-15705-0_10

by either direct or indirect means. PII is defined as information that directly identi-fies an individual (e.g., name, address, social security number or other identifying number or code, telephone number, email address, etc.) or by which a specific indi-vidual can be identified in conjunction with other data elements, that is; by indirect identification. These data elements may include a combination of gender, race, birth date, geographic indicator, and other descriptors [1]. This brings data privacy questions to the forefront. How secure is the data that is generated by IoT devices? How is it used? What happens to that data once the processing is complete? IoT data privacy is key [2].

In addressing data privacy regulations globally, particularly those required by the EU's General Data Protection Regulation (GDPR) that went into effect in May of 2018, the amount of data generated by the growing IoT is a major concern [3]. IoT device consumers, developers, manufacturers and government should be held responsible for their use of personal data.

10.2   IoT Security Challenges

Prospective consumers, investors and innovators of IoT-centric applications ponder these germane thoughts: IoT privacy, IoT usefulness, IoT effectiveness and trust-worthiness; with *Privacy* being critical. Internet of Things extends to everyday items not normally considered computers, allowing them to generate, exchange and consume data with minimal human intervention [4].

The main security challenges in IoT can be represented in relationship to the Information Security triad de facto standard of Confidentiality, Integrity and Availability (CIA). It can be argued that IoT security has emboldened the CIA triad with an increased focus on the security challenges of privacy, authenticity, and nonrepudiation, each challenge directly related to a specific leg of the CIA triad (Fig. 10.1).

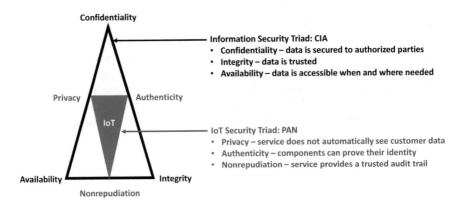

Fig. 10.1 Information Security Triad with IoT Security Triad

- Privacy is important for the data and the user. The data needs to be kept private at rest and in motion, in a device and on a network, in databases and the cloud. IoT systems desperately need privacy solutions that rely on strong security practices and privacy-by-design [4].
- Authenticity of the devices and individuals using or accessing the devices is tied to their identity, and the validation of their identity. It can be validated through a combination of technologies and processes. Mechanisms such as a root of trust (ROT) sometimes called a "Birth Certificate" embedded when the device is manufactured such as a Trusted Platform Module (TPM) or hardware based Physically Unclonable Function (PUF) could be deployed to enable authenticity [5].
- Nonrepudiation is an essential characteristic in all IoT scenarios which require trustworthy communications [6]. Nonrepudiation is the assurance that the validity of something cannot be denied, referring to proof of the origin and audit trail of data and the integrity of the data.

10.3   Privacy Management by Regulations

Privacy is complex and personal; an individual's perception of what privacy is varies. Privacy is crucial to protect and support the numerous freedoms and responsibilities that are germane to a democracy. The laws of society are the principal method of ensuring that the ability to freely exercise one's rights are protected; and that there are recourses when basic human rights are stripped away from a civic member.

Unfortunately, technological advancements have reached a point where the law cannot keep up with innovations and the constant changes technology brings to our lives. One key prevalent advancement is Big Data. Big Data is a term used across many industries and use cases to describe the large amount of data in the networked, digitized, sensor-laden, information-driven world [7]. Companies are leveraging Big Data and the Internet of Things to collect millions of facts about customers and using those facts to predict trends and develop better sales and marketing strategies. A store could claim that the technology is simply providing ways to better serve its customers; however, the store's clandestine purpose is to influence spending decisions by analyzing the sometimes private information they gather about their customers. Business entities are interested in influencing consumers' decisions, and so they learn as much about them as possible.

As we consider the privacy requirement domains in the world of Big Data and the Internet of Things, Fig. 10.2 represents an IoT system with a Wireless Sensor Network (WSN), networked with a processing, storage and analytics cloud; there are privacy requirements across these environments. In the Wireless Sensor Network (WSN) cloud there is a plethora of IoT devices (e.g., Internet-enabled security cameras). The processing, storage, and analytics cloud could be a public cloud such as Amazon Web Services, or a private cloud owned by an enterprise. Privacy management is relevant across these domains.

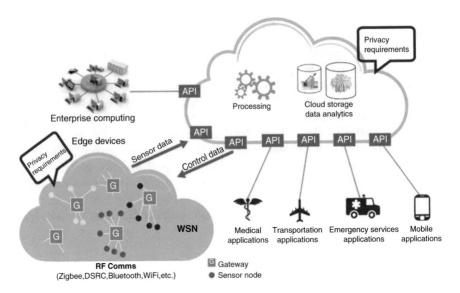

Fig. 10.2 Privacy requirement domains in an IoT system

10.4 IoT Privacy and The Law

Consumers of IoT must have confidence that their data collection, storage and usage is being done in a manner that benefits them and does not jeopardize their privacy. Reducing risks and building trust is essential regardless of an individuals' concerns about their privacy. Although not specifically passed for IoT, several applicable guidelines and regulations already addressed privacy, such as the Fair Information Practice Principles (FIPPs) [8], and specifically for European citizens the General Data Protection Regulation (GDPR) [3].

Complying with these principles and regulations requires a good understanding of the privacy risks in IoT systems. The privacy risks of a system are the product of three inputs: the personal data collected or generated, data actions performed on that information, and the context surrounding the collection, generation, processing, disclosure and retention of this personal data [9]. Reducing the privacy risk of an IoT system means reducing risks within all three dimensions. As a result, the GDPR sets "necessity" and "data minimization" as requirements to process personal data. For data actions and context, the regulation seeks consumer agreement via the principles of "transparency and openness," "notice," and "consent" [4]. Privacy solutions need to help consumers decide who should legitimately access and alter information. Given the general challenges of IoT, such as scale, dynamic changes, device heterogeneity, and resource-constrained IoT devices, some major risk factors are at play. These factors are discussed using three dimensions of risk in IoT—personal data, data actions, and context.

10.4.1 Personal Data: Consumers—Owners by Definition

According to GDPR, personal data means any information relating to "an identified or identifiable natural person; directly or indirectly, by a name, an identification number, location data, an online identifier or one or more factors specific to the physical, physiological, genetic, mental, economic, cultural or social identity" [3].

According to Ofcom, the UK communications regulator [10], personal data is:

- Volunteered data that comes directly from the individual.
- Observed data that is created as a result of a transaction between an individual and an organization.
- Inferred data, also called derived data, that is the output of data analysis, combination or mining.

To support consumers to take ownership of their personal data, IoT systems need to be transparent. Consumers should be able to review when their personal data is collected and how it is used and be able to give or withdraw consent [8]. More importantly, this consent should not only be "in or out" but be granular, that is, consumers should be able to choose a subset of data they would like to share with an IoT system. As a consequence, IoT systems should have the means to cope with missing data.

In addition, security solutions need to ensure confidentiality, integrity and availability of personal data. This, for instance, requires encryption when data is in transit and at rest, which may be challenging to support at the resource-constrained IoT devices [4].

10.4.2 Data Actions: Allowable Actions

GDPR [3] requires that consumers who are nationals of the EU must give consent "by a clear affirmative act establishing a freely given, specific, informed and unambiguous indication" of agreement to process personal data. Consent should cover all processing activities on such personal data, and "silence, pre-ticked boxes or inactivity" cannot constitute consent. Consent should also be as easy to withdraw as to give.

Therefore, IoT systems that handle personal data need to implement consent. Because IoT is a myriad of heterogenous systems, applications, networks and devices having different capabilities and varying technology implementations, controlling actions in IoT would require extending authorization to each segment of the IoT system as shown in Fig. 10.2. Authorization that supports the entire IoT system would need to be on the device or via a resource server that is outside of the device, and connects the device to the IoT system. Determining where to begin with authorization is challenging. IoT devices present higher risks, as they are resource-constrained and highly connected within an IoT system including other

IoT devices and the cloud. IoT systems may need to deploy more than one authorization solution to address requirements in the various subsystems of an IoT system.

10.4.3 Context: Achieving Context-Aware Privacy

Privacy in an IoT system is context-dependent and hence authorization should also be context-dependent. An appropriate access control method that can represent context is Attribute-Based Access Control (ABAC). In ABAC, a subject's request to perform operations on objects is granted or denied based on attributes of the subject and the object, environment conditions, and a set of policies that are specified in terms of those attributes and conditions. High-granularity policies are possible, for instance, using eXtensible Access Control Markup Language (XACML), describing exactly who interacts with the data, when, how, and to what extent [4]. The usage of privacy dashboards, notices, and recommendation systems is an effective way to guide and prod consumers to impact and take charge of their privacy.

10.5 IoT Privacy Management and You: Owning Your Privacy

The way governments view privacy, and the laws and regulations that govern privacy are important to understanding your own rights. Although it seems that every day fewer people care about their privacy, the ability to maintain parts of our life as private remains crucial to our democracy, our economy, and our personal well-being. However, we address the importance of maintaining your choices for what you wish to keep private. Your home, your body, your thoughts and beliefs are all within the control of the owner, and they are easier to hold private. Your finances, your relationships, and your sexuality are areas that most of us would consider private, although additional parties—your bank, your best friend, your sexual partner—hold information concerning these private matters, so privacy is expected, though absolute control is not possible. You may travel places on the public streets and therefore not expect absolute privacy, but you still expect to be relatively anonymous either in a crowd or a place where no one knows you. In this case, you would lose a measure of independence if everyone knew you everywhere you went and could tie together information about this trip with other related information they know about your shopping habits, your family history, and whose company you enjoy. Once your movements in space are recorded and added into the general knowledge-base without your permission, your privacy is limited. With the pervasive technology of IoT, loss of anonymity is rapidly increasing and the basic loss of the ability to keep secrets is in jeopardy. Data privacy protects consumers from retailer advertising, marketing, and promised customer incentives for participation in data collection and data mining of personal trends, patterns and preferences [11].

Loss of privacy can lead to coercion on personal choices, such as with the subtle pull of a retailer's website with "similar products bought frequently together" recommendations after the customer makes a purchase. For example, when a customer buys a winter boot, and after receiving the order confirmation e-mail, their e-mail and web browser are inundated with numerous recommendations of leg warmers "frequently bought together" with the boot they have just ordered.

10.5.1 Social Media and Social Engineering

Thieves, Scammers and Fraudsters target our privacy, particularly on social media. The more criminals (cyber-criminals) know about your money, your possessions, your travel habits, your security, and your vulnerabilities, the easier it will be to steal your privacy.

Choosing to post all your information on the Internet in Facebook, Twitter, Instagram, Snapchat, etc., or opting to tell everyone about your private business, makes you more vulnerable to many types of theft and scams. Choosing more privacy online can guard against all types of scam and privacy breaches. Setting your social media posts to specific privacy levels and audiences can help to restrict the viewers of your social media postings and activities, to help protect you from cybercriminals.

10.5.2 Location, Location: Enabling Your Privacy

Because privacy is a subjective and changing concept, your rights to privacy depend on where you live. For example, in the European Union (EU) and Canada, governments have established that it is the human right of every citizen to maintain direct control of businesses' and government's use of their personal data. Both jurisdictions have created privacy-protection forces that regulate the way personal data is collected and shared. Though regulations may be effective at protecting some data from use in certain business and government settings, they cannot stop people from blurting out information on social media. Many other countries, such as Israel, Switzerland, and Japan, have solid data protection regimes based on privacy protected as a human right. Other countries, such as India and Mexico, have protective laws in place but may not have a mature enforcement infrastructure to truly protect their people as Canada can. Conversely, the US federal government only protects certain classes of personal information. Unfortunately, the USA does not take the position that the ability to direct how business and government use personal information is a human right.

In the USA, state laws protect against exposure of customer data due to systems being hacked. However, these laws are inconsistent and usually are only relevant after the data has been lost because they address how a business must notify customers, patients, or employees when data has been exposed. In short, while many countries in the world protect private information in many ways, you still must be vigilant to protect your own private data. Even in Canada and the European Union, much of the information that you voluntarily expose through social media and in other media is beyond the government's protection. But in the USA and other countries, the private information that is unknowingly provided to business and government is not necessarily protected by law, and even US constitutional protections only assure citizens that a certain process will be undertaken before their lives can be interrupted by surveillance. While the government may sympathize with your need for privacy, no government will protect you as well as you can protect yourself [11].

10.6   Big Data Reigns

The deeper technology becomes embedded into our lives, the more it threatens our privacy. Technology, such as location trackers that are built into every smartphone and new car being sold today, allows a new window into our routines that was not available before. There was virtually no way to follow your regular movements until you started carrying and driving computers that reported location data. Sometimes the simple fact that we are using technology creates information that was never available before. For example, when you open a browser and sign onto the Internet, you are creating a type of record of your thoughts and actions that simply did not exist 25 years ago. When you sit on your couch and shop for Christmas gifts, watch fish videos on YouTube, check the weather in Washington DC for your trip, and then search for a recipe for homemade Hollandaise sauce, you have just created insights into your personality, travel destinations, eating and shopping habits, that no one would have been able to collect prior to the Internet's pervasive acceptance.

10.7   Conclusion

The fines for non-compliance with personal data regulations can be millions of dollars/euros, so it is essential that IoT device manufacturers, as well as those that use the devices, take the time to understand these regulations, and then consult with attorneys on an approach to personal data use, transfer, and storage. IoT data privacy needs to be built into these devices from the ground up, so that personal information remains secure [2]. Ultimately, consumers are responsible for their own privacy and maintaining control of the data they generate. How the data of the Internet of Things is used, stored, transmitted and accessed will require that consumers become more educated and informed about the technological advancements that they adopt and embrace, and the implications that the adoption has on their privacy. In countries such as the USA, consumers are at the helm of steering the "data privacy ocean-liner" into the channel that enables and protects their privacy.

References

1. https://www.dol.gov/general/ppii
2. https://innovationatwork.ieee.org/iot-data-privacy/
3. European Parliament and Council (2016) General Data Protection Regulation. Official Journal of the European Union, Brussels. https://eur-lex.europa.eu/legal-content/EN/ALL/?uri=CELE X%3A32016R0679
4. https://iot.ieee.org/newsletter/september-2016/
5. Network World (2018) Identifying the Internet of Things – one device at a time. https://www.networkworld.com/article/3287927/internet-of-things/identifying-the-internet-of-things-one-device-at-a-time.html
6. Oriwoh E, Al-Khateeb H, Conrad M (2015) Responsibility and non-repudiation in resource-constrained Internet of Things scenarios. https://doi.org/10.13140/RG.2.1.4030.3124
7. NIST (National Institute of Standards and Technology), NIST Big Data Interoperability Framework: Volume 1, Definitions, Special Publication (NIST SP) - 1500-1. https://www.nist.gov/publications/nist-big-data-interoperability-framework-volume-1-definitions
8. Gellman, "Fair Information Processing Practices," 2012. https://www.researchgate.net/publication/279297894_Responsibility_and_Non-repudiation_in_resource-constrained_Internet_of_Things_scenarios
9. NIST (National Institute of Standards and Technology) (2014) Privacy Engineering Objectives and Risk Model. Kantara Initiative IoT workshop
10. OFCOM (2015) Promoting investment and innovation in the Internet of Things. OFCOM, London
11. Payton T, Claypoole T (2014) Privacy in the age of big data: recognizing threats, defending your rights, and protecting your family. Rowman & Littlefield Publishers, Lanham. ProQuest Ebook Central, https://www.proquest.com/libraries/academic/

Chapter 11
Securing IoT Data with Pervasive Encryption

Eysha Shirrine Powers

11.1 Introduction

The Internet of Things (IoT) is considered one of the next great waves in computing. IoT is made up of network-connected devices and appliances equipped with digital sensors and microchips which are accessible through the internet. Many devices, such as smartphones and media players, are designed for built-in connectivity. Other traditional devices, such as thermostats and lighting systems, must be retrofitted for IoT connectivity.

Consider the following:

- A heart monitor which measures heart beats and alerts a doctor to abnormalities.
- Vehicles with built-in location tracking beacons so the vehicle can be located if stolen.
- Physical activity trackers, such as Fitbit, which measure the number of steps that you take in a day and track your progress towards your weight management goals.
- Video doorbells, such as Ring, which detect motion and alert users to visitors or intruders.
- Pet/child/elderly monitors that are motion or sound activated and allow two-way communication.
- Self-driving cars which navigate to users' target destinations using GPS and collision detection sensors.
- Scooters which have location beacons allowing users to pick them up anywhere and leave them anywhere.

E. S. Powers (✉)
Agoura Hills, CA, USA
e-mail: eysha@us.ibm.com

© Springer Nature Switzerland AG 2019
F. D. Hudson (ed.), *Women Securing the Future with TIPPSS for IoT*, Women in Engineering and Science, https://doi.org/10.1007/978-3-030-15705-0_11

In addition to consumer applications, IoT also applies to commercial and industrial environments. Thus, the possibilities for collecting, storing, processing, and sharing IoT data are endless.

11.2   Why Secure IoT Data?

Consider medical data in the United States which is subject to data privacy and security regulations. Specifically, the Health Information Portability and Accountability Act (HIPAA) of 1996 and the Health Information Technology for Economic and Clinical Health Act (HITECH) of 2009 require safeguards for personal health information. Personal health information, known as "individually identifiable health information," includes the following:

- The individual's past, present, or future physical or mental health condition.
- The provision of health care to the individual.
- The past, present, or future payment for the provision of health care to the individual.

The HIPAA Privacy Rule protects all "individually identifiable health information" held or transmitted by a covered entity or its business associate, in any form or media, whether electronic, paper, or oral [1]. This is considered "protected health information" (PHI). Covered entities include health insurance companies, Health Maintenance Organizations (HMOs), and government sponsored programs like Medicare and Medicaid as well as health care providers like doctors, clinics, dentists, psychologists, chiropractors and pharmacies. With the adoption of HITECH, HIPAA regulations also apply to business associates which handle health data on behalf of covered entities such as transcriptionists and claims processors.

The HIPAA Security Rule establishes national standards to protect PHI that is created, received, used, or maintained by a covered entity or its business associate. It requires administrative, physical, and technical safeguards to ensure the confidentiality, integrity, and security of electronic protected health information [2].

Medical data collected from IoT devices are subject to HIPAA and HITECH regulations. Payment data collected from IoT devices are subject to the Payment Card Industry Data Security Standard (PCI-DSS). European Union (EU) citizen data collected from IoT devices are subject to the EU General Data Protection Regulation (GDPR). Whether owing to industry or government regulations, or the threat of a data breach, IoT data must be properly secured.

11.3   The Information Life Cycle

To understand the complexity of securing IoT data, we must understand the information life cycle. Security experts consider the information life cycle to include the acquisition, use, archival, and disposal of data (Fig. 11.1).

Fig. 11.1 Information Life
Cycle

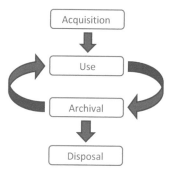

Information moves between the various phases as follows:

- Information acquisition occurs when information is created, generated and/or copied.
- Information use occurs when information is read or modified.
- Information archival occurs when data is deemed no longer in use but must be retained for regulatory, backup or other reasons. Note that archived information can be recalled to active use.
- Information disposal occurs when data is destroyed.

We can apply the information life cycle to health data in an IoT environment. For example, sensors in a heart monitor detect electrical signals for each heartbeat [3]. The data is recorded in the IoT device and automatically transmitted to a healthcare provider. The healthcare provider receives the health information and structures the data. The data in its final form can be stored in a database, or other data store, for use in diagnosis by the healthcare provider. When the data is no longer in use, it may be archived and ultimately disposed.

During its life cycle, IoT data may reside in the following:

- The physical IoT device.
- The internet packet transmitted to the healthcare provider.
- Memory of the receiving application on the provider's server.
- A database which writes the data to a file or data set.
- Active disk or tape storage.
- Archived storage which may or may not be offsite.
- A disaster recovery backup system.

A pervasive encryption strategy can be developed and implemented to safeguard sensitive data in an IoT environment. Data in-transit can be protected with network encryption. Data at-rest can be protected with application-level, database-level, file or data set-level, and/or disk and tape-level encryption. The use of pervasive encryption can strengthen an organization's security posture and protect sensitive IoT data from hackers and malicious insiders.

11.4   How Encryption Works

Encryption is a major component of cryptography. Cryptography (or "crypto" for short) is defined as the practice and study of techniques for securing information in the presence of malicious third-parties known as adversaries. Encryption, using an algorithm and a key to transform an input into an encrypted output, specifically prevents the disclosure of information to unauthorized individuals. In an IoT environment where sensitive information must be protected, encryption can be employed to ensure that only authorized users have access to safeguarded data.

Encryption operations require a cryptographic key, input text and an algorithm. The output of the encryption operation is the encrypted data which is known as ciphertext. The ciphertext can be decrypted using the appropriate cryptographic key to produce the original input text or plaintext. If the cryptographic key is not compromised, and the algorithm is not vulnerable, the ciphertext is secure wherever it resides.

11.4.1   Cryptographic Keys

Cryptographic keys are the main source of security for encrypted data. If the key is easy to guess or easy to access, then the encrypted data is at risk. Therefore, secure creation and management of cryptographic keys are critical to a pervasive encryption strategy.

The two basic types of cryptographic keys are symmetric keys and asymmetric keys. Symmetric keys enable the bulk encryption of data. A single symmetric key both encrypts and decrypts data. Asymmetric keys enable secure distribution of symmetric keys and support digital signatures. Asymmetric keys are generated in pairs. The asymmetric public key encrypts data and verifies digital signatures. The asymmetric private key decrypts data and generates digital signatures.

11.4.1.1   Symmetric Keys

Symmetric keys are random numbers typically represented as bytes. A byte is an 8-bit value. A bit can have only two values, 0 or 1. Therefore, a value represented in bits (or bytes) is known as a binary value. Bits can be grouped together and represented by hexadecimal, which may be prefixed with "0x", or other number systems. For example, the 8-bit value "11101001" is the same as the two-digit hexadecimal value "0xE9". In this case, a single byte (8 bits) is represented by 2 hexadecimal digits. Binary values are often represented in hexadecimal so that they can be represented with fewer digits.

The following example shows a 256-bit random number represented in binary and hexadecimal. The binary representation has 256 digits, where each digit represents one bit. The hexadecimal representation has 64 digits, where each digit represents four bits.

Binary

1010101111001000001010101110000010000110101011100010001101111111
1111010101100001111011111111010100100100001001000111100001010111101
1100010011010111001100010110110110011110000100111101000010001001
11100110000110010110000001011100100000111000110000100101001 10

Hexadecimal

A B C 8 2 A E 0 8 6 A E 2 3 7 F E A C 3 D F D 4 9 0 9 1 E 1 5 E E 2 6 B 9 8 B
76784F4227CC32C0B907184A6

The length of the symmetric key (i.e., random number) depends on the cryptographic algorithm. For example, the Advanced Encryption Standard (AES) algorithm supports 128-bit, 192-bit, and 256-bit key lengths. The Data Encryption Standard (DES) supports 56-bit, 112-bit, and 168-bit key lengths. Standards bodies, such as the National Institute of Standards and Technology (NIST), recommend the use of long keys for encryption. Long keys are less susceptible to brute force attacks, in which the attacker tries to guess the correct key.

Brute force attacks on symmetric keys typically involve exhausting every possible combination of bits until the encrypted data is decrypted into readable text. For example, consider a single-length DES key, which has 56 bits. Since there are two possible bit values (0 or 1), there are $2^{56} = 72,057,594,037,927,936$ or 72 quadrillion possible combinations of bits. A brute force attack involves trying each one of those combinations until the correct key is discovered. When the DES algorithm was invented in the 1970s, the cost of purchasing or building a computer with the speed to break a DES key was so high that it was deemed infeasible. Today, you can purchase a customized computer for less than $250,000 which can break a DES key in 3 days [4].

11.4.1.2   Asymmetric Keys

Asymmetric keys are not merely random bit strings. Generating asymmetric key pairs involves complex math operations. With asymmetric keys, one key can be distributed publicly as a public key and one key can be kept private as a private key. Anyone can encrypt data with the public key but only the owner of the private key can decrypt the data. The concept behind the generation of asymmetric keys resembles a trap door, where it is easy to go through a trap door one way but very difficult to go back the other way. So the owner of the private key can easily generate the public key. However, the user of the public key cannot generate the owner's private key.

The length of the asymmetric key depends on the cryptographic algorithm. Typical key lengths for the Rivest–Shamir–Adleman (RSA) algorithm are 512-bit, 1024-bit, 2048-bit, and 4096-bit. The Elliptic Curve algorithms defined by NIST (i.e., NIST PRIME curves) are P-192, P-224, P-256, P-384, and P-521. Due to the complexity of the math operations, the time to generate asymmetric keys rises significantly as the key lengths increase.

The following table shows the performance for asymmetric key generation, using an IBM Crypto Express adapter [5] (Table 11.1).

Table 11.1 Performance of
asymmetric key generation

Algorithm and key length	Operations/s
RSA CRT 512-bit	106
RSA CRT 1024-bit	79.1
RSA CRT 2048-bit	20.5
RSA CRT 4096-bit	1.89
EC P-192	1158
EC P-256	809
EC P-521	307

Brute force attacks on asymmetric keys are more difficult due to the need to solve the "hard problems" underpinning the trap door functions. For example, prime factorization is considered a "hard problem." A prime number is a whole number greater than 1 which can only be divided evenly by 1 and itself. We can define a prime number, p, with a value of 521 and another prime number, q, with a value of 719. Multiplying p and q results in a number, n with a value of 374,599. It is easy to take prime numbers p and q, and multiply them together to produce n. It is also easy to factor a small number, n, into its prime factors p and q. However, as n becomes large, the effort increases in factoring it into prime values p and q. Prime factorization is the underlying "hard problem" which makes brute force attacks on RSA keys difficult. The number 374599 can be represented as the 20-bit number '01011011011101000111'b which would be considered the modulus of a 20-bit RSA key. In 2015, researchers developed a method to factor a 512-bit RSA key in under 4 h for $75 [6]. As computing speed increases, key lengths must be increased to ensure that factorization remains difficult.

11.4.2 Input Text

Encryption input can be any clear text such as "Hello World." However, encryption operations are performed on binary data. Thus, standard characters must be converted to binary data (i.e., 0 s and 1 s) using encodings. An encoding defines a numeric value for every character.

American Standard Code for Information Interchange (ASCII) encoding is a standard for electronic communications. The ASCII encoding for "Hello World" in hexadecimal is "0x48656c6c6f20576f726c64". The capital letter "H" is represented by "0x48". The lowercase letter "e" is represented by "0x65". The lowercase letter "l" is represented by "0x6C". The space is represented by "0x20". Each character whether it is uppercase, lowercase, numeric or special, can be represented in binary.

Extended Binary Coded Decimal Interchange Code (EBCDIC) encoding is the standard encoding for IBM mainframes. The EBCDIC encoding for "Hello World" in hexadecimal is "0xC88593939640E696999384". The capital letter "H" is represented by "0xC8". The lowercase letter "e" is represented by "0x85". The lowercase letter "l" is represented by "0x93". The space is represented by "0x40". The binary form of the input text is used in the math operations defined by the encryption algorithm.

11.4.3   Output Text

Ciphertext is the expected output of encryption. Ciphertext is typically represented in binary (although there are new methodologies available such as Format Preserving Encryption, FPE). The length of the ciphertext output depends on the length of the clear text input and encryption algorithm used to produce the ciphertext. The ciphertext length will always be greater than or equal to the input length.

11.4.4   Algorithm

Mathematical operations performed by an encryption algorithm vary based on the type of encryption to be performed, the length of the input data, and the size of the encryption key. Some of the most widely used encryption algorithms today are the Data Encryption Standard (DES), Advanced Encryption Standard (AES), Rivest–Shamir–Adleman (RSA), and Elliptic Curve Cryptography (ECC). DES and AES are symmetric algorithms. RSA and ECC are asymmetric algorithms.

Symmetric algorithms can operate on blocks of data or streams of data. AES and DES algorithms support block ciphers, which operate on blocks of data. Specifically, AES supports 16-byte blocks. Therefore:

- If there are 16 characters (i.e., 16 bytes) of input text, AES would successfully encrypt the entire block.
- If there are less than 16 characters (i.e., 16 bytes) of input text, the input text must be padded to fill an entire block.
- If there are more than 16 characters (i.e., 16 bytes) of input text, the input text must be divided into 16-byte blocks.
  - The mode of operation indicates whether the blocks will be operated on independently as in Electronic CodeBook mode (ECB), or if the output of one block will be combined into the input of another as in Cipher Block Chaining mode (CBC) or another chaining scheme.
  - If the final block is less than 16 bytes, then it must be padded to fill the block.

Asymmetric algorithms operate on data in its entirety. There is no blocking, chaining or input padding concept that is applicable to asymmetric algorithms. For the RSA algorithm, the input data size is limited to the size of the modulus. For example, a 4096-bit RSA key is one of the largest RSA key lengths. Since a byte is 8 bits, a 4096-bit key amounts to 512 bytes. An ASCII character is the equivalent of one byte. Therefore, a 4096-bit RSA key can only encrypt 512 characters of input text. Of course, there are many sources of text that are more than 512 characters. Since the length of the input text is limited, asymmetric encryption is not typically used for encrypting standard text. Instead, it is used for encrypting small pieces of data such as symmetric encryption keys.

11.5 Pervasive Encryption

Pervasive Encryption is a strategy in which data is encrypted everywhere it travels and/or resides in the computing environment. Pervasive encryption rests on existing technologies such as encryption and existing infrastructure including cryptographic hardware to enable the fast, secure, and reliable protection of data throughout its life cycle.

In our IoT example, sensors in a heart monitor detect electrical signals for each heartbeat. The data is recorded in the IoT device and automatically transmitted to a healthcare provider. The healthcare provider receives the health information and structures the data. The data in its final form can be stored in a database or other data store for use in diagnosis by the healthcare provider. When the data is no longer in use, it may be archived and ultimately disposed. The data must be protected at each of these points.

11.5.1 Data in Transit

Sensitive information collected from sensors in IoT devices is typically transmitted electronically to backend computer systems for processing and storage. As the data travels through the internet to the provider's server, the data is susceptible to disclosure. Encryption can be used to ensure the data remains secret as it traverses external and internal networks.

11.5.1.1 Network Encryption

Network encryption provides a means of ensuring data remains secure as it travels over the network to its destination. A connection protocol can be used to ensure that communications between an IoT device and the server are secure. One example of a connection protocol is a handshake. The handshake protocol begins when the IoT device connects to the server. The device initiates a secure connection and sends along information about its crypto capabilities. The server chooses the strongest crypto options and responds with its public key in a certificate to identify itself. The IoT device validates the server against a certificate authority before using the public key. If the server certificate and public key are valid, the IoT device generates a symmetric key and encrypts the symmetric key with the server's public key. The encrypted key is sent across the network to the server. The server uses its private key to decrypt the symmetric key. Now, the IoT device and server share a symmetric key. The IoT device can encrypt and send sensitive data across the network to a validated server which only that server can decrypt.

11.5.1.2 Coupling Facility Encryption

Coupling Facility (CF) encryption is a unique concept for enterprise systems built on mainframes. Coupling facilities operate as the hub in a star topology for communications between connected mainframes. Mainframes can cache and/or share IoT data between systems that are connected to the coupling facility. Sensitive data transmitted through and/or residing in the coupling facility may be encrypted using CF encryption (Fig. 11.2).

CF Encryption is configured by policy. For each CF structure enabled for encryption, the coupling facility invokes a crypto library to generate a 256-bit AES key which is encrypted by a Master Key (MK). The encrypted key is stored in the Coupling Facility Resource Management (CFRM) data set.

When CF encryption is enabled, data is encrypted on the host (e.g., SYS1) prior to transmission to the CF structure. The encrypted data flows as follows:

1. On SYS1, the AES key is retrieved from the CFRM data set.
2. On SYS1, the data is encrypted by the AES key.
3. As the data flows to the CF structure, it remains encrypted.
4. As the data sits in the CF structure, it remains encrypted.
5. On SYS3, the AES key is retrieved from the CFRM data set.
6. On SYS3, data is requested from the CF structure.
7. As the data flows from the CF structure to SYS3, it remains encrypted.
8. On SYS3, data is decrypted using the AES key (Fig. 11.3).

Using CF encryption, data is protected end-to-end as it travels between mainframe systems and environments.

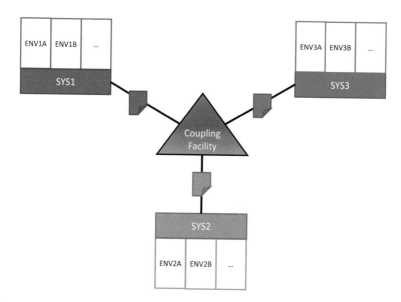

Fig. 11.2 Coupling Facility (CF)

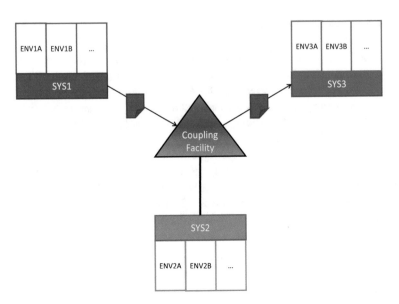

Fig. 11.3 Coupling Facility (CF) Encryption of data sent from SYS1 to SYS3

11.5.2 Data at Rest

Sensitive information collected from sensors in IoT devices that has been transmitted to one or more computing environments will reside in computer memory and hardware storage such as hard disks and tape drives in various locations. The data becomes accessible to system administrators, database administrators, storage administrators, security administrators, data center technicians, and computer hackers. While access controls, such as READ, WRITE, or NONE authority, can be used to control access to sensitive data, encryption can provide an additional layer of protection.

11.5.2.1 Application Level Encryption

Sensitive IoT data can be sent to an application to classify and structure the information before storing the data in a database. If the finalized data is stored in the database unencrypted, a database administrator (DBA) would have the authority to view the sensitive data in the clear, unencrypted. The DBA requires UPDATE or higher authorization to perform their day-to-day tasks, however, the elevated authorization implicitly permits READ authorization to the data. If the data is encrypted in the application before it is written to the database, the DBA would only see the ciphertext in the database. Therefore, the data is protected from view.

Encrypting data at the application level is considered application level encryption. Data encrypted at the application level remains encrypted as it is written to a database, written to a file or data set, and written to a hard disk or tape drive.

Application level encryption requires deep knowledge of cryptography, crypto coding skills, and application changes if algorithms become vulnerable or out of date.

11.5.2.2 Database Level Encryption

Sensitive IoT data can be encrypted at the database level as an alternative or in addition to application level encryption. Database level encryption is much simpler to implement than application level encryption. Implementation typically involves a configuration file or data set which indicates the type of symmetric encryption (e.g., AES or DES) and the cryptographic key or keys. Many organizations use database level encryption when they lack the resources and/or skills to do application level encryption.

Database level encryption supports encryption of data in rows or columns. When a request to read an encrypted field is received, the data must be decrypted so that it is readable to the user. Thus, the encryption of database fields can slow down day-to-day database operations. In environments where availability is a concern, high-speed encryption hardware or alternative encryption methods might be advantageous.

11.5.2.3 File and Data Set Level Encryption

Sensitive IoT data can be encrypted at the file and/or data set level. The concept of data set level encryption is new and unique to mainframe environments. Data set encryption is implemented by policy. For each data set resource, a new field is supported to indicate that newly created data sets should be encrypted with a given key. To encrypt existing data sets, new data sets must be allocated, and the original data must be copied to the encryption enabled data sets.

Mainframe data set level encryption uses 256-bit AES keys for encryption. When data is read or written to a data set, it is decrypted or encrypted with the associated key as it is accessed. Consequently, there are two controls around the data. One access control authorizes users to perform actions on the data set. A separate access control authorizes users to the cryptographic key that encrypts and decrypts the data. Therefore, a storage administrator with UPDATE access to the dataset but NONE access to the key can create, rename, archive and delete a data set but cannot read from or write to the data set.

11.5.2.4 Disk and Tape Level Encryption

Sensitive IoT data can be encrypted at the hard disk and tape drive level as well. In fact, most enterprise environments utilize disk and tape encryption. Disk and tape encryption is the simplest level of encryption to configure and maintain. When enabled, all data on the disk and tape device is protected with encryption.

Disk and tape level encryption uses symmetric encryption for the data. The key sizes vary based on the hardware device and/or encryption software. Disk and tape level encryption is designed for offline attacks. For example, if a data center technician has physical access to the disks and tapes, they can remove them from the machine and read them elsewhere. When disk and tape level encryption is in place, physical removal of encrypted media yields unreadable data. For other users, such as storage administrators and system administrators, the data is decrypted as it leaves the disk, and thus it is readable in the clear if it is not encrypted using another method.

11.6   Choosing Encryption Engines

The successful deployment of a pervasive encryption strategy, which employs one or more methods of encrypting data in transit, and/or one or more methods of encrypting data at rest, hinges on the speed and reliability of the cryptographic engines performing the encryption operations. The underlying crypto engines must support symmetric key generation, encryption and decryption as well as asymmetric equivalents. In computing environments with heavy workloads, the cryptographic engines must be able to keep business operations performing within acceptable response times. At the same time, the cryptographic engines must also meet regulatory requirements for security and compliance.

11.6.1   Hardware Engines

Hardware cryptographic engines may be designed for speed and/or security. These include hardware security modules (HSMs) which may be Peripheral Component Interconnect Express (PCIe)-attached or network-attached, and CPU-integrated cryptographic hardware.

11.6.1.1   Hardware Security Modules

HSMs are cryptographic modules which typically have tamper responding enclosures and are designed to meet higher level security certifications such as FIPS 140-2 and ISO 19790:2012 [7]. Tamper responding enclosures have intrusion detection mechanisms. For example, IBM Crypto Express adapters, respond to temperature thresholds. When tampered, the adapters erase all configuration data and clear all cryptographic keys and registers. IBM Crypto Express adapters are also designed to ensure that neither sound emissions nor electronic discharges reveal subtle information about the cryptographic operations being performed.

Table 11.2 Performance comparison of crypto hardware [5]

Data Length (bytes)	Operations/s (CPU)	Operations/s (HSM)
64	875,252	10,569
256	871,629	9126
1024	714,697	9019
4096	462,037	8225
64 K	52,915	1017
1 M	3521	67.64

11.6.1.2 CPU-Integrated Cryptographic Hardware

CPU-integrated cryptographic hardware has the benefit of residing side by side with data processing circuits, resulting in a lower latency for performing the crypto operations. The speed in which crypto operations can be performed in this environment is a major advantage over HSMs.

In a mainframe environment, CPU-Integrated cryptographic hardware can provide up to 80x faster performance than Hardware Security Modules. Table 11.2 shows a comparison of the number of AES-CBC encryptions per second that can be performed using a 256-bit key with CPU-Integrated cryptographic hardware such as the IBM Central Processor Assist for Cryptographic Function (CPACF) versus a Hardware Security Module such as the IBM Crypto Express 6S.

11.6.2 Software Engines

Software cryptographic engines may be utilized when dedicated hardware is unavailable. These engines can perform simple or complex cryptographic operations. For example, Java programming libraries support the Java Cryptography Extension (JCE). The software JCE provider can run on computer hardware supporting the Java Virtual Machine (JVM). Java Card, which is tailored to smart cards and memory-constrained devices, supports a subset of JCE functions which includes key generation and encryption. Although software cryptographic engines traditionally have a slower execution time, they enable the use of cryptography in a wide array of devices including small IoT devices.

11.7 Managing Encryption Keys

Encryption keys are the core of encryption operations and the management of those encryption keys should be the core of a pervasive encryption strategy. When data is encrypted, the availability and integrity of the encryption keys impact the readability of data which may be critical to an organization. As such, there must be considerable forethought and planning for managing the encryption keys.

In our IoT example, we could use network encryption to encrypt sensitive data in transit from the heart monitor to the server. We could use application level encryption to encrypt the data before it is stored in a database. We could use database encryption to protect specific health fields in the tables. We could then use data set level encryption to encrypt the data sets underlying the database. We could also use disk encryption to protect data from offline attacks.

For each area where encryption is utilized, there must be a key management process. Each organization is unique. Questions such as the following must be addressed:

- Who owns the keys? Who creates the keys? Who are the key custodians? How many keys will be generated? How are keys requested?
- Where will the keys be stored? Where will the keys be backed up? How often will the keys be backed up? How will the keys be mapped to the data they are protecting?
- Who has access to the keys? Who should not have access to the keys?
- How often do keys need to be rotated? Which keys will be rotated?
- When will keys be archived? Will keys ever be deleted? How can you be sure a key is no longer needed?
- How is key corruption detected and recovered? How are accidentally deleted keys recovered?

11.7.1 Key Ownership

Determining key ownership, creation and custodians depends on the size of the organization and data ownership. Some organizations will grant key ownership to the data owners. In this way, the data owners decide when keys need to be created and who will have access to the keys. Some organizations will manage keys at a system level. The system administrator owns the keys and ensures system-wide resources are protected.

11.7.2 Key Storage

Encryption keys are typically stored in a key store. Key stores can exist in several forms. Java supports password protected key store files. SSL supports key database files. The IBM z/OS operating system supports key data sets. Keys are identified in key stores by a label or handle and/or other attributes of the key. The naming convention for the keys should be determined such that the key owners or custodians can easily identify which keys protect which data. Some key stores, such as IBM's key data sets, support custom metadata which can be used to add additional information about a key such as the key owner's email address. One or more backup processes are essential regardless of where the keys are stored.

11.7.2.1 Physical Backup

Physical backups are a standard process for most computing environments. They typically involve an automated backup system which takes backups of folders, drives or volumes on a regularly scheduled basis (e.g., daily or weekly). With respect to key management, physical backups may not be enough. For example, if a physical backup is taken on a Friday and hundreds of keys are generated on Monday which are used for encrypting sensitive data on Tuesday, but the volume containing the key store is corrupted on Wednesday, then the Friday backup would not be able to restore the system to its working state. There must be a procedure in place to ensure that the system can be restored to a working state as quickly as possible.

11.7.2.2 Logical Backup

Logical backups are backups that are taken before or after a major change. For example, prior to generating hundreds of new keys, a key custodian may backup the previous key store. If a mistake is made during key generation, the previous key store can be restored. The same key custodian may make a backup immediately after the new keys are generated. If the active key store file becomes corrupted, the new keys have not been lost and can be restored quickly from the backup. Logical backups are critical for key management.

11.7.2.3 Offline Backup

Offline backups are necessary to ensure malicious attacks or accidental corruption are not unrecoverable. An offline backup is a backup which is not connected to the network. It may be a backup that is written to tape or other removable media. The idea is that corruption (or sabotage) of data on the active system would not impact the offline media. Therefore, the system could be restored to a working state. Offline backups are not the same as an offsite backup. Even an offsite backup can become corrupted if data is automatically copied to that location. A disadvantage of offline backup is that the data may not be as current.

11.7.3 Key Protection

Cryptographic keys are sensitive resources which must be protected. Three ways to protect encryption keys are access controls, key wrapping and separation of duties. Access controls are the standard method of protecting encryption keys. Authorization to encryption keys and/or key stores should be granted based on a business need. Storage administrators may have authorization to files or data sets containing encryption keys. In that case, key wrapping can protect the keys at rest.

Key wrapping involves encrypting the key with another key. So, a storage administrator can read the data set containing the keys but cannot use the keys because the key values are encrypted with a key that is not stored in a file or data set. On IBM Z systems, we call that key-encrypting key a Master Key. The Master Key is loaded onto a Crypto Express adapter. As a result, the Master Key is inaccessible to the host (i.e., operating system) environment. Separation of duties is often recommended to provide additional protection around the Master Key. Separation of duties would involve generating multiple master key parts. Each master key part would have a different custodian (or key owner). All master key parts must be combined to activate the Master Key. Thus, no individual person has access to the complete key.

11.7.4   Key Rotation

Key rotation is a process in which a piece of data is decrypted with an old key and encrypted with a new key. In some cases, key rotation is also considered the process where existing data is encrypted with an existing key in a given period (e.g., a month), and new data will be encrypted with a new key in the next period. The piece of data could be health information, or in the case of key wrapping, a key itself. Organizations must stay abreast of various industry regulations and their own security policies which will have guidelines for rotating keys.

Key compromise is when a key has been stolen by a malicious party. Key rotation is an absolute imperative in the case of a key compromise. If a key has been compromised, any data which was encrypted with that key must be re-encrypted with a new key. The compromised key may be monitored, archived and/or destroyed after ensuring all data has been re-encrypted.

11.7.5   Key Archival

Key archival is a recommended alternative to key deletion. In an environment where data may exist in various computer systems, key custodians must consider whether non-compromised keys should ever be destroyed. Generally, the life of the key must align with the life of the data protected by that key. For example, in an environment where data set encryption is implemented, the life of the key is the life of the data set. If the data set contains health information, then that data set may need to exist for the lifetime of an individual which could be up to 100+ years. In that 100-year span, sensitive data may exist in databases, files, data sets, disk drives, and/or tape drives. If the encryption key is destroyed, all encrypted data is lost along with it.

11.7.6 Key Availability

Key availability is important for disaster recovery. An offsite disaster recovery (DR) environment should be in place in case of a disaster at the primary site. To process encrypted data, the DR site must have a configuration similar to the primary environment. The key stores must be available. The crypto hardware must be installed, online and configured. Crypto software and middleware must be installed and configured. The DR procedure for ensuring that the cryptographic infrastructure required for data encryption and decryption must be tested and validated.

11.8 New Technologies on the Horizon

Cryptographic technology continues to evolve. As attackers become more creative in discovering algorithm weaknesses and vulnerabilities to exploit, researchers become more inventive in finding new ways to protect data.

11.8.1 Homomorphic Encryption

Homomorphic encryption is a new cryptographic technology in research. Homomorphic encryption is designed to provide a means to perform mathematical operations on encrypted data which when decrypted, will result in the correct result. IBM research, led by Craig Gentry, invented a working algorithm to perform homomorphic encryption. The challenge in using those coding libraries in real world applications is performance. Encryption operations must be high-speed to ensure that normal business processes can run steadily as sensitive data is being protected. Today's homomorphic encryption operations are found to be significantly slower than similar plaintext operations [8]. Research teams will continue to look for ways to improve homomorphic encryption performance for practical use.

11.8.2 Crypto Anchors

Crypto anchors are a blockchain-based technology that can validate the authenticity of objects such as pharmaceuticals. The research is being led by IBM's Andreas Kind [9]. Crypto anchors, which are smaller than a grain of salt, are designed to be embedded in physical objects which can be traced back through their supply chain to the originator. Counterfeit objects would be easily distinguished from genuine objects. Crypto anchors are underpinned by cryptographic technology and is envisioned to be one of the five innovations that will help change our lives in 5 years [10].

11.9 Conclusion

Encryption technology is supported in small devices and large computing environments. IoT data begins its life on the IoT device and moves across networks into applications, databases, files, and storage media. Protecting sensitive data wherever it exists ensures that organizations can meet industry regulations, such as HIPAA, while safeguarding data against malicious users and data breaches.

A pervasive encryption strategy necessitates a choice of strong algorithms and key sizes that stand against brute force attacks. Encrypting data in applications, databases, and files requires a robust cryptography infrastructure. Cryptographic engines may need to support high-performance transactions and high-security standards such as ISO 19790 and FIPS 140-2. Procedures for key management and disaster recovery must be well-defined and rehearsed prior to deployment.

The Internet of Things will continue to expand and the data which is communicated between various computing systems and which resides in various locations must be safeguarded. How will you protect it?

Acknowledgments Thank you to Kenneth Kerr, Eleanor Chan, and Todd Arnold at IBM for reviewing the technical content and examples.

References

1. U.S. Department of Health & Human Services (2003) Summary of the HIPAA Privacy Rule. Retrieved from U.S. Department of Health & Human Services: https://www.hhs.gov/sites/default/files/ocr/privacy/hipaa/understanding/summary/privacysummary.pdf
2. Office of the Secretary (2003) 45 CFR Parts 160, 162, and 164. Retrieved from Department of Health and Human Services: https://www.hhs.gov/sites/default/files/ocr/privacy/hipaa/administrative/securityrule/securityrulepdf.pdf?language=es
3. Johns Hopkins Medical (n.d.) Event Monitor. Retrieved from Johns Hopkins Medical: https://www.hopkinsmedicine.org/healthlibrary/test_procedures/cardiovascular/event_monitor_135,333
4. Electronic Frontier Foundation (2016) EFF DES cracker machine brings honesty to crypto debate. Retrieved from Electronic Frontier Foundation: https://www.eff.org/press/releases/eff-des-cracker-machine-brings-honesty-crypto-debate
5. IBM (2017) IBM z14 Performance of Cryptographic Operations. Retrieved from IBM: https://www-01.ibm.com/common/ssi/cgi-bin/ssialias?htmlfid=ZSL03607USEN
6. Factoring as a Service (2015) Retrieved from International Association for Cryptologic Research: https://eprint.iacr.org/2015/1000.pdf
7. International Organization for Standardization (2018) ISO/IEC 19790:2012. Retrieved from International Organization for Standardization: https://www.iso.org/standard/52906.html
8. Chirgwin R (2018) IBM's homomorphic encryption accelerated to run 75 times faster. Retrieved from The Register: https://www.theregister.co.uk/2018/03/08/ibm_faster_homomorphic_encryption/
9. IBM (n.d.) Crypto-anchors and Blockchain. Retrieved from IBM: https://www.research.ibm.com/5-in-5/crypto-anchors-and-blockchain/
10. IBM (2018) Five innovations that will help change our lives within five years. Retrieved from IBM: https://www.research.ibm.com/5-in-5/

Chapter 12
Secure Distributed Storage for the Internet of Things

Sinjoni Mukhopadhyay

12.1 Outline

With recent advances in machine-to-machine communication built on cloud computing and networks of data-gathering sensors, the Internet of Things (IoT) has become an important part of the system and software community. IoT devices generate massive amounts of data which makes the cloud an ideal storage solution for such data. Although cloud storage may be preferred over on-premise storage due to its ease of access, low cost and delegated infrastructure management, IoT data storage using a single cloud provider has several disadvantages. Relying on a single cloud provider to store all the data compromises data reliability while simultaneously sacrificing data availability. IoT devices generate different types of data, all of which are efficiently organized using different storage solutions. This chapter discusses leveraging distributed storage properties to securely store different types of IoT data while simultaneously ensuring data availability.

A few years ago, the idea of placing workloads on a single public or private cloud seemed very enticing; but since the introduction of hybrid cloud architectures [1], the choices in terms of the variety of services available has made it a more attractive option for many enterprises. As more enterprises are avoiding dependency on a single public cloud provider, cloud computing is making a shift towards the multi-cloud strategy. The main difference between multi-cloud and hybrid cloud solutions, in this author's opinion, is the manner in which the resources are deployed in each model. Hybrid cloud solutions can consist of a combination of private and public cloud deployments, all a part of a single cloud service provider, or multiple private or public cloud deployments; whereas multi-cloud solutions can include either private or public cloud deployments belonging to different cloud service pro-

---

S. Mukhopadhyay (✉)
Computer Science, University of California, Santa Cruz, CA, USA
e-mail: simukhop@ucsc.edu

© Springer Nature Switzerland AG 2019
F. D. Hudson (ed.), *Women Securing the Future with TIPPSS for IoT*, Women in Engineering and Science, https://doi.org/10.1007/978-3-030-15705-0_12

viders. Additionally, all functions in a hybrid cloud are performed as a conjoint effort between public and private clouds, whereas in multi-cloud solutions different services are provided by different cloud vendors. Using a multi-cloud strategy has a certain degree of flexibility which allows the user to choose between a variety of features made available at the most competitive pricing. Organizations also believe that a multi-cloud strategy has other benefits like the following: it helps avoid vendor lock-ins; it tackles cloud reliability; it allows organizations to pick between services that are best suited to their applications/workload, and in addition it also provides benefits of data sovereignty. Data sovereignty defines the idea that information which has been stored in binary/digital form is subjected to the laws of the country in which it is located. The multi-cloud strategy allows organizations to leverage geographically dispersed clouds to meet data sovereignty requirements and improve the user experience.

Distributed cloud storage allows data to be stored across multiple storage servers, either belonging to the same cloud or belonging to different cloud providers. With data being distributed across servers, the following questions arise:

1. How can this distributed data be recovered?
2. How easily can an adversary regroup data chunks to steal the original data?
3. How is data recovered if a subgroup of servers is failing?

This chapter discusses one such distributed storage technique that has major reliability and availability advantages. After introducing the distributed storage technique, we go on to talk about how this technique answers each of these questions.

Secret-sharing is a distributed storage algorithm used to store data chunks across multiple servers. Each server has an erasure coded piece of the data, which reveals no information about the original data. Data can be recovered as long as a sufficient number of servers are functional and available. Additionally, in case of data breaches in a single server, the attacker only gets access to the piece of the data stored under that server, which does not provide sufficient information to generate the original data object. This chapter outlines the existing storage solutions for different types of IoT data. It outlines the limitations of the existing state of the art, while addressing the benefits of applying a secret-sharing based IoT data model.

12.2 Background

Mark Hung, vice-president of Gartner Research stated that "The Internet of Things will have a great impact on the economy by transforming many enterprises into digital businesses and facilitating new business models, improving efficiency and increasing employee and customer engagement [2]." Based on reports provided by Gartner analysts, illustrated in Table 12.1 [3], looking into the future, there will be a 33.7% Compound Annual Growth Rate (CAGR) in units of IoT devices or connected things, from 6.4 billion in 2016 to 20.4 billion in 2020, an overall increase of 220%.

Table 12.1 IoT installed Base by Category (Millions of Units) and CAGR (Compound Annual Growth Rate) [3]

Category	2016	2017	2018	2020	CAGR (%)
Consumer	3963.0	5244.3	7036.3	12,863.0	34.2
Business: Cross-Industry	1102.1	1501.0	2132.6	4381.4	41.2
Business: Vertical-Specific	1316.6	1635.4	2027.7	3171	24.6
Grand Total	6381.8	8380.6	11,196.6	20,415.4	33.7

Table 12.2 Predicted rise in endpoint spending and CAGR (Millions of Dollars) [3]

Category	2016	2017	2018	2020	CAGR (%)
Consumer	532,515	725,696	985,348	1,494,466	29.4
Business: Cross-Industry	212,069	280,059	372,989	567,659	27.9
Business: Vertical-Specific	634,921	683,817	736,543	863,662	8.0
Grand Total	1,379,505	1,689,572	2,094,881	2,925,787	20.7

The number is projected to rise beyond 2020, with an increased number of consumers purchasing more IoT devices, and businesses spending more to develop and maintain such devices. In 2017, in terms of hardware spending, the use of connected things among businesses was expected to drive 964 billion dollars in total across the cross-industry and vertical-specific business segments as shown in Table 12.2.

IoT Endpoint Spending for Consumer applications was expected to amount to 985 billion dollars in 2018. By the end of 2020, hardware spending from both segments is projected to reach nearly 3 trillion dollars. This predicted rise in the Internet of Things business will pave the way and define a mandate for optimizations of newer system and security technologies to make the future IoT devices more efficient and secure. We will next explore the secure storage and cloud technologies which may be leveraged to meet this mandate.

12.2.1   IoT Storage in the Cloud

The importance of data storage decisions stems from their implications in terms of application performance, data integrity, and data protection and restoration. The strategic decision to move from on-premise storage to cloud storage can be based on the initial costs, maintenance, type and amount of storage, that is, the Total Cost of Ownership (TCO) [4]. There are two main criteria that need to be kept in mind while selecting a storage solution, that is, security and cost. The nature of your data determines the best storage location for this data, such that this data can be accessed securely as required by the business without violating security and regulatory needs. The cost can be determined based on a detailed consumer needs (like expected latency and data availability) versus vendor study. This would include the analysis of needs of different workloads and databases and their matched capabilities and costs with different cloud vendors.

In terms of system performance, hardware and software, system upgrades in the Cloud are much faster and inexpensive as compared to the months needed to upgrade on-premise storage. Disaster recovery in clouds is much more reliable in terms of the backup instances that are replicated in various physical locations, as compared to on-premise storage in tapes and disk backups. However, when it comes to security, technologists are conflicted between choices. The Snowden breach [5] in 2013 and the more recent Amazon Web Services (AWS) outage [6] in 2017 are two examples that prove that data is not secure whether on-premise or on the cloud.

The large volume and heterogenous types of data generated by the Internet of Things make the Cloud, with its high processing power, an ideal solution that some enterprises choose to process this data rather than build huge amounts of in-house capacity. Some examples of current cloud giants tackling IoT workloads are Microsoft's Azure IoT solution accelerators [7], Amazon's AWS IoT [8] and Google's Cloud IoT [9]. In general, cloud computing resources are fairly inexpensive in terms of availability and can also perform tasks rapidly. They easily adapt to the needs of each user that they serve, and the user location is generally irrelevant to the usage of the cloud from a technology perspective, but there sometimes could be regulatory implications. As long as you have the Internet, you can connect to the devices.

Currently, both IoT devices and the Cloud need designers and programmers to fix various incompatibilities to make sure they can work together better in the future. Figure 12.1 shows the layers involved with IoT data storage in the cloud. In the

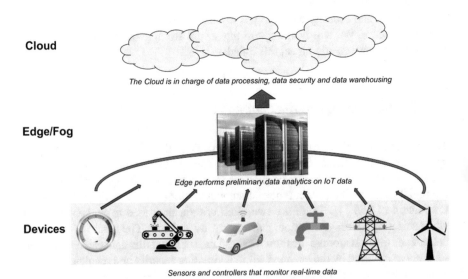

Cloud

The Cloud is in charge of data processing, data security and data warehousing

Edge/Fog

Edge performs preliminary data analytics on IoT data

Devices

Sensors and controllers that monitor real-time data

Fig. 12.1 The Cloud is an efficient storage solution for the massive amounts of data generated by the Internet of Things [10, 11]

future, some IoT devices will have increased device power to improve on-device data processing. Taking care of data processing and management on the device-level would mean that the Cloud resources can focus on things like data security, availability and storage. On the other hand, the Cloud deployments need to securely transfer and store data that are being communicated from different devices in different locations. In order to combine the flexibility and high computation power of the cloud with the intelligence of the IoT devices, new ways of analyzing and storing data is on the rise. This mitigates the dependency on cloud storage and enhances the role of IoT in performing computations and making decisions while staying closer to the end-user.

This new way is Edge [12] and Fog [13] computing. Instead of sending raw data off to the Cloud to be disseminated and analyzed, edge/fog computing is used for devices that require instantaneous or real-time decision making from their sensors in order for them to function correctly. An example of such a device is self-driving cars, which will create a new subsection of machine-to-machine communication in the form of vehicle-to-vehicle (V2V) communication. Any interactions with these cars will need to happen as close to real-time as possible. Edge computing moves data a far shorter distance, as compared to cloud, from the sensors themselves to local gateway device such as a switch or a router. This gateway device then performs the necessary processes and analysis and sends back decisions to the IoT device quicker than via cloud computing. Typically, this is done by the IoT devices transferring the data to a local device that includes compute, storage and network connectivity in a small form factor. Data is then processed at the edge, and all or a portion of it is sent to the central processing or storage repository in a corporate data center, co-location facility or Infrastructure as a Service (IaaS) cloud. Edge computing has various use cases. One is when IoT devices have poor connectivity and it is not efficient for IoT devices to be constantly connected to a central cloud. Other use cases involve latency-sensitive processing of information. Edge computing reduces latency because data does not have to traverse over a network to a data center or cloud for processing, making it ideal for situations where latencies of milliseconds can be untenable, such as in financial services or manufacturing. Fog computing is a term derived from edge computing. Fog refers to the network connections between edge devices and the cloud. Edge, on the other hand, refers more specifically to the computational processes being done close to the edge devices.

12.2.2 Security in the Internet of Things

The IoT industry at this time is in its infancy and has significant security and privacy implications. Matt Burgess mentions, "Everything that's connected to the internet can be hacked, IoT products are no exception to this unwritten rule." He further goes on to talk about the toy manufacturer Vtech, which lost the videos and pictures of children due to hackers compromising its insecure IoT systems [14]. Earlier, Wikileaks claimed that the CIA has been developing security exploits for a connected Samsung television. The US director of national intelligence, James Clapper, predicted in 2016 that "In the future, intelligence services might use the Internet of Things for identification, surveillance, monitoring, location, tracking, and targeting for recruitment, or to gain access to networks or user credentials" [15].

In general, security is one of the biggest issues with the Internet of Things. Data collected by sensors may be extremely sensitive; for example, what we say or do at our own house, what we like and dislike, and our interests, to name a few. Security of such information is vital to consumer trust. But based on the track record provided by IoT devices, basic security concepts have been given little thought. The lack of patching capability in most IoT devices makes it hard to recover from software flaws detected on a regular basis. Hackers have actively begun targeting IoT devices such as routers and webcams because the inherent lack of security in these devices makes them easy to compromise and roll up into giant botnets. In 2017, researchers found 100,000 webcams originating in China that could be hacked with ease [16]. Some of these NeoCoolCam devices were priced as low as 39 dollars and were purchased all around the world. They contained improper quality assurance at the firmware level, several bugs affecting their authentication mechanisms and other buffer overflow vulnerabilities [17]. In 2017, other IoT vulnerabilities around the world have been brought to light like the internet-connected smartwatches for children, investigated by the Norwegian Consumer Council, that have been found to contain security vulnerabilities that allow hackers to track the wearer's location, eavesdrop on conversations, or even communicate with the user [18]. Currently, the tradeoff between cost and security has made the abovementioned problems widespread and intractable.

The Internet of Things bridges the gap between the digital world and the physical world, which means that hacking into devices can have dangerous real-world consequences. For example; hacking into the sensors controlling the temperature in a power station could trick the operators into making a catastrophic decision about incorrectly modulating the temperature even when not needed; taking control of a driverless car could lead to major accidents and risks of losing lives. In short, people should understand that there are many different use cases for the Internet of Things, many of which are yet to be explored, and that this has the potential to positively and negatively impact our lives. This chapter aims to throw light on some of the use cases and explore potential solutions to the problems we may face in the future due to the inherent properties of the Internet of Things.

12.3 Storage and the Internet of Things

IoT devices typically have limited data storage capabilities. Most of the data needs to be communicated using protocols such as Message Queuing Telemetry Transport (MQTT) [19] or Constrained Application Protocols (CoAP) [20], then further ingested by IoT services for additional processing and storage. MQTT is designed for connections with remote locations or where network bandwidth is limited, whereas CoAP is used for constrained devices that communicate within the same constrained network. Dealing with the increased volume of data has made it difficult to secure the data in storage and to maintain integrity and privacy of the data. Apart from the obvious "more data means more storage" problem, there is also the added problem of dealing with different types of data generated by these devices. First, there is large-file data, such as images captured from medical devices. This data type is typically accessed sequentially. The second data type is very small, for example, log-file data captured by sensors. These sensors, while small in size, can create billions of files that must be accessed randomly. Determining the type of data to be stored is an essential first step to finding an optimized storage solution. The final goal is to build a multi-tiered storage solution that will work well with all types of data.

12.3.1 Existing Storage Technologies

Based on previously explained challenges with IoT data, storage solutions need to have three main properties: they need to securely store massive amounts of data, support horizontal scaling, and need to have the ability to deal with heterogeneous data resources. This section describes some existing storage technologies that are currently being applied to the Internet of Things.

Fazio et al. propose a two-layer hybrid architecture based on both SQL-like, and XML-like non-SQL technologies to provide a scalable, efficient and elastic sensing service [21]. They represent heterogeneous monitoring devices and data using Sensor Web Enablement (SWE) specifications, which defines data encodings and web services to store and access sensor-related data. Kang et al. propose a sensor-integrated radio frequency identification (RFID) data repository-implementation model using the MongoDB database [22]. They use a design based on horizontal data partitioning to maximize query speed and uniform data distribution over data servers. Gray et al. proposed a low complexity greedy distributed data replication mechanism to increase resilience and storage capacity of IoT based surveillance systems against node failure and local memory shortage [23]. Hu et al. introduce a ubiquitous data accessing method to deal with distributed storage of IoT based data in the healthcare area [24]. Teing et al. perform a forensic investigation of peer to peer (P2P) cloud storage for IoT networks using BitTorrent as a case study outlining strengths and weaknesses of using P2P cloud for IoT networks [25].

Jiang et al. talk about a database management model that combines multiple databases to store and manage structured IoT data [26]. They also propose a Representational State Transfer RESTful service generating mechanism to provide a hypertext transfer protocol (HTTP) interface for those applications that access the data stored based on their framework. Liu et al. propose a storage management solution based on NoSQL called IOTMDB [27]. Apart from handling large-scale data, this solution also tackles data sharing and collaboration. They also provide a query mechanism to easily search and locate their shared data. Raj et al. go in a different direction as compared to literature around the time and exploit a document-oriented approach to propose a system that supports both heterogeneous and multimedia data [28]. They built their storage solution on top of the CouchDB database server and used a RESTful API to provide a rich set of features that targeted generic IoT applications. Distributed storage techniques came into the picture long before systems engineers started using secret- sharing to store their data.

Shyu et al. modify Shamir's secret-sharing [29] to utilize all coefficients in polynomials for larger data capacity at the data level [30]. Additionally, they use a distributed IoT storage infrastructure to provide scalability and reliability at the system level. Multiple IoT storage servers are aggregated to improve storage capacity, whereas individual servers can join and leave freely for flexibility at the system level. The importance of securely storing data is increasing as the amount of data produced by IoT devices is increasing. Satarkar et al. propose a secure storage architecture for IoT data, that combines Rivest–Shamir–Adleman (RSA) and AES to encrypt different files, of different sizes and contents [31]. Jararweh et al. simplify IoT management by proposing a software-defined based framework model to securely store data produced by IoT objects [32]. Shafagh et al. use blockchain, as an auditable and distributed access control layer on top of the storage layer, to enable secure and resilient access control management [33]. Their system accommodates for IoT data streams where streams are chunked, compressed, and encrypted in the application layer and authorized services are granted access only to the decryption keys.

12.4   Secret-Sharing

First invented by Adi Shamir and George Blakley in 1979, secret-sharing refers to all methods that can be used to distribute a secret among multiple participants in a group, such that each participant in the group only has a share of the secret and not the entire secret. The secret can be regenerated with the sufficient number of shares and individual shares are meaningless on their own. Figure 12.2 shows an overview of the secret-sharing technique. The regeneration of the secret is determined by a threshold scheme (T, N), where N is the total number of shares in the secret and T is the sufficient number of shares needed to regenerate the secret. Secret-sharing can be used as an alternative to traditional encryption techniques that tradeoff between confidentiality and reliability. The choice associated with storing an encryption key securely is either keeping a single copy of the key in one location for maximum confidentiality or having multiple copies of the key in different locations for maximum reliability. A single copy of the key for confidentiality allows for a single point of compromise and stealing that key would result in a storage system breach. Secret-sharing allows administrators to avoid key management issues. Having multiple copies of the key for reliability would allow an adversary to steal any of the copies which would lead to a breach of the storage system.

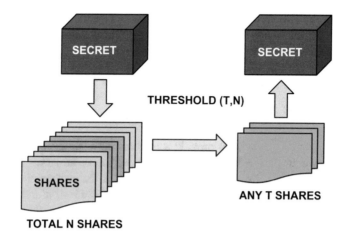

Fig. 12.2 Shamir's secret-splitting generates N equal-sized object shares out of which any T shares are both necessary and sufficient to rebuild the original object

12.4.1   Math Behind Secret-Sharing

The secret-sharing problem arises when a secret requires a certain number of participants to be mandatorily present before it can be revealed. Imagine a dying person who wants to leave his estate and wealth to his five children. He wants to make sure that the four older children do not end up bullying the youngest child to avoid giving him his share of riches. Therefore, the person makes a will stating their shares and locks it up. He splits the key into five parts such that the will can only be accessed when all five children are present and willing to unlock the box. We assume the will to be an integer S, that is the secret, the number of participants N or 5 in the case of this example and a minimum number of participants required for reconstruction T or 5 in the case of this example. The minimum number of participants required for reconstruction can be less than the value of N.

"The Mathematics of Secret Sharing" defines the math through a distributing algorithm and the reconstruction algorithm [34]. The distributing algorithm D accepts inputs S, N, T and produces an output list of N numbers $D(S, N) = \{x_1, x_2,..., x_N\}$. These represent the secrets distributed to the N participants. The reconstruction algorithm R accepts T numbers $\{y_1,..., y_T\}$ as input and gives an output number M. These algorithms are designed to hold two main properties:

- Knowledge of T or more shares makes S easily computable.
- Knowledge of less than T shares leaves S completely undetermined.

The main idea behind the secret sharing protocol is a polynomial interpolation. Polynomial interpolation states that given $k + 1$ points on a plane with x distinct values, there is a unique degree k polynomial. As a byproduct of this statement, there are infinitely many degree $k + 1$ polynomials that pass through the same points. The proof that such a polynomial definitely exists is divided into two parts, proof of existence for the polynomial and proof of uniqueness for the polynomial. Refer to article "Proof of existence and uniqueness using splitting fields" for mathematical proofs of uniqueness and existence [35].

12.4.2 Types of Secret-Sharing

Secret-sharing has been built to provide a tradeoff between security and performance. The two main subclasses of secret-sharing are information-theoretically secure and computationally secure secret-sharing. Information-theoretically secure is when any number of shares less than the defined threshold is insufficient to generate the original data. The limitation of this type of secret-sharing is that each object share is of the same size, which means that the storage and transmission bandwidth required by the shares is equivalent to the size of the secret times the number of shares. In computationally secure secret sharing, shares are a fraction of a size of the secret. This uses repeated polynomial interpolation making the computations more complex and time-consuming, therefore allowing for data to be more secure. This can be used for secure information dispersal on the Web and in sensor networks.

Trivial secret-sharing has three types with threshold $T = 1$, $T = N$ and $1 < T \leq N$. For $T = 1$ trivial secret-sharing, the secret can be distributed to all N participants. For $T = N$ trivial secret-sharing, all shares of the secret are needed to reconstruct the secret. Trivial secret-sharing for $1 < T \leq N$ is where the complexity begins as we have to construct a secure secret-sharing scheme without needing all shares of the secret to rebuild the secret. Shamir came up with an information-theoretically secure secret-sharing scheme that uses Lagrange Interpolation and the size of each share does not exceed the size of the secret. The advantage of Shamir's secret-sharing is that keeping T constant, shares can easily be added or removed without affecting other shares. Proactive secret-sharing allows users to change threshold number with every update of the system but expects them to keep track of malicious users keeping expired shares. The verifiable secret-sharing scheme guarantees users that the other users in the group are not concealing or lying about the contents of their shares. Any of the above-mentioned types of secret-sharing can be used based on the type of data we choose to secure.

12.4.3 Applications of Secret-Sharing

Secret-sharing has been used as a method of secure information dispersal for a lot of different types of data. For example, for archival storage systems, it is preferable to use the information-theoretically secure secret-sharing over computationally-secure secret-sharing. This is because our adversary is assumed to have unlimited computation power and time, which means that given enough time he will eventually be able to compute the secret. However, a short-term data storage system can be secured by just using computationally secure secret-sharing.

Apart from archival storage secret-sharing has had many other applications (e.g., storing multimedia data). Shyu et al. talk about parallel implementations of Shamir's threshold secret-sharing scheme using sequential Central Processing Unit (CPU) and parallel Graphics Processing Unit (GPU) platforms to show that GPU can achieve a lucrative speedup over CPU when dealing with shared multimedia data [29]. Roy et al. have used (k, n) image secret sharing that involves sharing of a secret image into n number of pieces called shadow images in such a way that k or a greater number of shares can retrieve the original image [36]. They have proposed a (3, 4) image secret sharing scheme that has adopted the concept of visual cryptography over the 2X2 block. Security of the system is enhanced by scrambling the blocks using a pseudo-random sequence. Chen and Wu introduce a secure Boolean-based secret image sharing scheme which uses a random image generating function to generate a random image from secret images or shared images [37]. This efficiently increases the sharing capacity or storage bandwidth that is used to share the random image. Ching tackles Chen and Wu's inaccurate multi-secret image sharing (MSIS) by proposing a strong threshold (n, n) MSIS scheme without leaking partial secret information from $(n - 1)$ or fewer shared images. Komargodski et al. [38] construct a computational secret-sharing scheme for any monotone function in non-deterministic polynomial-time (NP) assuming witness encryption for NP and one-way functions. This results in a completeness theorem for secret-sharing where the computational secret-sharing scheme for any single monotone NP-complete function implies a computational secret-sharing scheme for every monotone function in NP. Huang et al. [39] propose secret sharing schemes to improve decoding bandwidth. Additionally, they consider the setting of secure distributed storage where the proposed communication efficient secret sharing schemes not only improve decoding bandwidth but further improve disk access complexity during decoding. Rawat et al. [40] talk about the centralized multi-node repair (CMR) model wherein multiple storage nodes can be reconstructed simultaneously at a centralized location, but there is a tradeoff between the amount of data stored and repair bandwidth. They provide another application of secret sharing in communication, where the codes for the multi-node repair problem are used to construct communication efficient secret sharing schemes with the property of bandwidth efficient share repair. Bai et al. propose a computationally secure and non-interactive verifiable secret sharing scheme that can be efficiently constructed from any monotone Boolean circuit [41]. Harn uses Lagrange's components, which are a linear combination of shares, to reconstruct a secret [42]. They extend their scheme to multi-secret sharing schemes as well. They compare existing multi-secret sharing schemes based on cryptographic assumptions like secure one-way function or solved discrete logarithm problem. Hadavi et al. present multiple partitioning methods that enable clients to efficiently search among shared secrets while preventing inference attacks on the part of data servers, even if they can observe shares and queries [43].

12.5   Secret-Sharing for the Internet of Things

Massive amounts of data from the Internet of Things are being stored on the cloud. Secret-sharing can be used with multiple deployments of the cloud to save cost while simultaneously improving the security of the data. Data can be grouped based on how frequently it is accessed or how secure it needs to be. The two kinds of deployments we talk about here are the hybrid cloud deployment and the multi-cloud deployment. Hybrid clouds can use a combination of private and public clouds to store all shares of data from IoT devices. Public clouds offer features such as scalability, low cost, and flexibility, while private clouds offer features like security, customization, enhanced control, and predictable costs. For example, if we were to look at an IoT network that consisted of devices that generated real-time data and data that needs to be stored long term, Figure 12.3 shows how the data may be distributed among multiple cloud instances.

For data that needs to be stored long-term, secret-sharing can be performed, and less than the threshold number of shares can be stored in a public cloud and any shares larger than that number can be stored in the private cloud. The main benefit of such a model would be that we can reduce the cost of storing an object by distributing shares between the private cloud and the cheaper public cloud. Additionally, our objects will be secured as part of the shares that are in the more secure private cloud. In case of a breach in the public cloud, shares stolen by an adversary would be less than the threshold number and therefore not useful in rebuilding the object (we are assuming here that private cloud is more secure than public cloud). A multi-cloud deployment can be used to prevent users from trusting a single cloud provider to store their data securely. Such a model could enhance data privacy by distributing object shares among multiple cloud providers, and may also be fault tolerant towards server downtimes or failures. Every provider would have shares of the data, but not a sufficient number of shares needed to rebuild the original data object. This model can be used to store any generated data that does not need to be immediately sent to the devices as feedback.

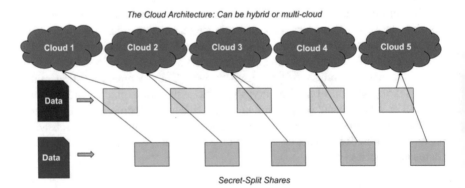

Fig. 12.3 IoT data storage using hybrid cloud or multi-cloud architectures

12.6 Conclusion

Storage systems for data produced by the Internet of Things need to provide properties like data security, availability, and scalability, and also be able to handle heterogeneous types of data. Cloud storage is an efficient way to store the massive amounts of data being generated by IoT devices as it has provisions for all the above properties. This chapter highlights issues with current storage techniques for IoT data and suggests secret-sharing as an efficient way to securely store such data on the Cloud. Secret-sharing is a distributed storage technique that not only stores data securely but also provides data availability. Additionally, the different types of secret-sharing techniques (information-theoretically secure and computationally secure) make it an ideal solution for both long- and short-term storage of data. In this chapter, we hypothesize about two secure distributed cloud storage models- a hybrid cloud model and a multi-cloud model, each of which has different advantages (cost versus availability). Both models may perform computations on gateway devices and therefore can work with IoT devices that have a wide range of computing power. Our proposed storage models could prove to be extremely efficient and secure for the recently rising IoT ecosystems used to run smart cities and smart hospitals.

References

1. Hybrid Cloud Architecture. https://aws.amazon.com/enterprise/hybrid/
2. Hung M. Leading the IoT, Gartner insights on how to lead in a connected world. https://www.gartner.com/imagesrv/books/iot/iotEbook_digital.pdf
3. Meulen R. Gartner Says 8.4 Billion Connected "Things" Will be in Use in 2017, Up 31 Percent From 2016: https://www.gartner.com/en/newsroom/press-releases/2017-02-07-gartner-says-8-billion-connected-things-will-be-in-use-in-2017-up-31-percent-from-2016
4. Scheier B. Decision guide: public cloud versus on-premise storage. https://www.hpe.com/us/en/insights/articles/decision-guide-public-cloud-versus-on-prem-storage-1701.html
5. Snowden E. Leaks that exposed US spy program. https://www.bbc.com/news/world-us-canada-23123964
6. Massive Amazon Cloud service outage disrupts sites. https://www.usatoday.com/story/tech/news/2017/02/28/amazons-cloud-service-goes-down-sites-scramble/98530914/
7. Azure IoT solution accelerators. https://azure.microsoft.com/en-us/features/iot-accelerators/
8. AWS IoT. https://aws.amazon.com/iot/
9. Google Cloud IoT. https://cloud.google.com/solutions/iot/
10. Mainframes and supercomputers, from the beginning till today. http://www.cpushack.com/2018/05/27/mainframes-and-supercomputers-from-the-beginning-till-today/
11. 4 Edge Computing Technologies Enabling IoT-Ready Network Infrastructure. https://www.lanner-america.com/blog/4-edge-computing-technologies-enabling-iot-ready-network-infrastructure/
12. Edge Computing. https://en.wikipedia.org/wiki/Edge_computing
13. Fog Computing. https://en.wikipedia.org/wiki/Fog_computing
14. Burgess M. What is the Internet of Things, WIRED explains. https://www.wired.co.uk/article/the Internet of Things-what-is-explained-iot
15. Timm T. US Intelligence Chief: We might use the Internet of Things to spy on you. https://www.theguardian.com/technology/2016/feb/09/the Internet of Things-smart-home-devices-government-surveillance-james-clapper

16. Ranger S. What is IoT? Everything you need to know about the Internet of Things right now. https://www.zdnet.com/article/what-is-the-the Internet of Things-everything-you-need-to-know-about-the-iot-right-now/
17. 175,000 IoT cameras can be remotely hacked thanks to flaws, says security researcher. https://www.zdnet.com/article/175000-iot-cameras-can-be-remotely-hacked-thanks-to-flaw-says-security-researcher/
18. Security flaws in children's smartwatches make them vulnerable to hackers. https://www.zdnet.com/article/security-flaws-in-childrens-smartwatches-make-them-vulnerable-to-hackers/
19. MQTT Wikipedia. https://en.wikipedia.org/wiki/MQTT
20. Constrained Application Protocol Wikipedia. https://en.wikipedia.org/wiki/Constrained_Application_Protocol
21. Fazio M, Celesti A, Villari M, Puliafito A (2014) The need of a hybrid storage approach for IoT in PaaS cloud federation, 28th international conference on advanced information networking and applications
22. Kang YS, Park IH, Rhee J, Lee YH (2016) MongoDB-based repository design for IoT-generated RFID sensor big data. IEEE Sensors J. https://doi.org/10.1109/JSEN.2015.2483499
23. Gray V, Gonizzi P, Ferrari G, Leguay J (2015) Data dissemination scheme for distributed storage for IoT observation systems at large scale. Inf Fusion. https://doi.org/10.1016/j.inffus.2013.04.003
24. Hu J, Xu B, Xu LD, Bu F (2014) Ubiquitous data accessing method in Iot-based information system for emergency medical services. IEEE Trans Ind Inform. https://doi.org/10.1109/TII.2014.2306382
25. Teing YY, Dehghantahna A, Yang L (2017) Forensic investigation of P2P cloud storage services and backbone for IoT networks: BitTorrent sync as a case study. J Comput Electr Eng. https://doi.org/10.1016/j.compeleceng.2016.08.020
26. Jiang Z, Jiang L, Xu LD, Xu B (2014) An IoT-oriented data storage framework in cloud computing platform. IEEE Trans Ind Inform. https://doi.org/10.1109/TII.2014.2306384
27. Liu Y, Li T, Mao W (2012) A storage solution for massive IoT data based on NoSQL. IEEE international conference on green computing and communications
28. Raj M, Francesco MD, Li N, Das SK (2012) A storage infrastructure for heterogeneous and multimedia data in the internet of things. IEEE international conference on green computing and communications
29. Shyu SJ, Tsai YZ (2018) Shamir's secret sharing scheme in parallel. In: Peng SL, Wang SJ, Balas V, Zhao M (eds) Security with intelligent computing and big-data services. SICBS 2017. Advances in Intelligent Systems and Computing (vol. 733). Springer, Cham
30. Chen V, Jiang H, Shen F, Jeong YJ (2015) A secure and scalable storage system for aggregate data in IoT. Future Gener Comput Syst. https://doi.org/10.1016/j.future.2014.11.009
31. Satarkar P, Bokefode J, Bhise A, Modani D (2016) Developing a secure cloud storage system for storing iot data by applying role-based encryption. Procedia Comput Sci. https://doi.org/10.1016/j.procs.2016.06.007
32. Jararweh Y, Al-Ayyoub M, Darabseh A, Benkhelifa E, Vouk M, Rindos A (2015) SDIoT: a software defined based Internet of Things framework. J Ambient Intell Humaniz Comput 6:453–461
33. Shafagh H, Burkhalter L, Hithnawi A, Duquennoy S (2017) Towards blockchain-based auditable storage and sharing of IoT data. Proceedings of the 2017 on cloud computing security work-shop
34. Chadha K. The mathematics of secret-sharing. https://jeremykun.com/2014/06/23/the-mathematics-of-secret-sharing/
35. NTNU. Proof of existence and uniqueness using splitting fields. https://wiki.math.ntnu.no/_media/ma3202/2015v/ch16-splittingfields.pdf
36. Roy R, Bandhopadhyay S, Kandar S, Dhara BC (2015) A novel 3–4 image secret sharing scheme. International conference on advances in computing, communications and informatics

37. Chen CC, Wu WJ (2014) A secure Boolean-based multi-secret image sharing scheme. J Syst Softw. https://doi.org/10.1016/j.jss.2014.01.001
38. Komargodski I, Naor M, Yogev E (2016) Secret-sharing for NP. J Cryptol 30:444–469
39. Huang W, Langberg M, Kliewer J, Bruck J (2016) Communication efficient secret sharing. IEEE Trans Inf Theory 62:7195–7206
40. Rawat AS, Koyluoglu OO, Vishwanathan S (2016) Centralized repair of multiple node failures with applications to communication efficient secret sharing. Cornell University Library
41. Bai G, Damgard I, Orlandi C, Xia Y (2016) Non-interactive verifiable secret sharing for monotone circuits. Springer International Conference on Cryptology
42. Harn L (2013) Secure secret reconstruction and multi-secret sharing schemes with unconditional security. Secur Commun Netw. https://doi.org/10.1002/sec.758
43. Hadavi MA, Jalili R, Damiani E, Cimato S (2015) Security and searchability in secret sharing-based data outsourcing. Int J Inf Secur 14:513–529

Chapter 13
Profiles of Women Securing the Future with TIPPSS for IoT

Florence D. Hudson

13.1 Introduction

This book celebrates the wise wonderful women who are working on securing the future through their work in TIPPSS—trust, identity, privacy, protection, safety and security—related to the Internet of Things [1–4]. They are technologists, astrophysicists, aerospace engineers, computer scientists, biochemists, cybersecurity professionals, experts in identity and access management (IAM), policy experts, lawyers, judges, students, and venture capitalists, with experience in industry, academia, and government. The majority have a doctorate degree. They are very accomplished, and impassioned to make the world a better, safer place. They are leaders in their field, and in their communities, in their countries, and across the world. As you read these women's stories, please think about how we can inspire more young girls and women to pursue their professional and personal passion, to be our leaders of tomorrow. Let them inspire you to join the world of TIPPSS and the Internet of Things, to make the world a better and safer place.

13.2 Author Profiles of Women Securing the Future with TIPPSS for IoT

Alicia D. Johnson is the resilience and recovery manager for the San Francisco Department of Emergency Management. Her work uses human-centered design principles to build collaborative relationships between the community and disaster responders to better protect the people and places we value. She has responded to

F. D. Hudson (✉)
Purchase, NY, USA
e-mail: Florence.distefano.hudson@gmail.com

© Springer Nature Switzerland AG 2019
F. D. Hudson (ed.), *Women Securing the Future with TIPPSS for IoT*, Women in Engineering and Science, https://doi.org/10.1007/978-3-030-15705-0_13

numerous disasters and large events including the 2008 Democratic National Convention, 2012 Superstorm Sandy, 2015 Supreme Court Ruling on Gay Marriage and San Francisco PRIDE Celebration, 2016 Super Bowl 50, 2017 UPS Active Shooter, and the 2017 Northern California Fires. She regularly serves as an Emergency Operations Center Manager for the City of San Francisco. Her work has inspired countless communities to educate their residents about disaster preparedness using connection rather than fear. She received her bachelor's degree in communications and political science and master's in Public Administration from the University of Colorado. She currently serves as a Senior Fellow for the West Big Data Innovation Hub [5]. When she is not in the Emergency Operations Center, Alicia can easily be found camera in hand, documenting the world around her [6, 7].

Britt Danneman is an investor at the Los Angeles-based venture capital firm Alpha Edison [8]. She primarily invests at the Seed to Series B stages to help set business model, fundamental strategy, team, and capital plan. She invests thematically, with one such focus on trust. Prior to joining Alpha Edison, Britt worked in distressed and middle market investing at Bain Capital Credit in Boston. She also worked in corporate development and strategy in San Francisco at the fintech startup Funding Circle. She received her MBA from Harvard Business School and undergraduate degree in Finance and Management from The Wharton School at The University of Pennsylvania [9].

Hon. Cynthia D. Mares, Esq., is a district court judge in the 18th Judicial District of Colorado. She is Senior Legal Advisor and Advisory Board Member for Axon Global [10], a cybersecurity company located in Houston, Texas. She is also Vice Chair for the Colorado Gaming Commission. Judge Mares is a 2016 alumna of the Harvard Kennedy Executive Education program and a governance fellow with the National Association of Corporate Directors since 2016. She is past president of the Hispanic National Bar Association and Colorado Hispanic Bar Association [11].

Edna Conway serves as Cisco's chief security officer, Global Value Chain, creating clear strategies to deliver secure operating models for the digital economy [12]. She has built new organizations delivering cyber security, compliance, risk management, sustainability and value chain transformation. She drives a comprehensive security architecture across Cisco's third-party ecosystem. Recognition of her industry leadership includes: membership in the Fortune Most Powerful Women community, a Fed 100 Award, Stevie "Maverick of the Year Award," a Connected World Magazine "Machine to Machine and IOT Trailblazer" Award, an SC Media Reboot Leadership Award, a New Hampshire TechProfessional of the Year Award, and CSO of the Year Award at RSA. She holds a JD from the University of Virginia School of Law, and a bachelor's degree from Columbia University, with executive education at Stanford University, MIT and Carnegie Mellon University. Prior to Cisco, she was a partner in an international private legal practice and served as Assistant Attorney General for the State of New Hampshire [13].

Eysha Shirrine Powers is a senior software engineer at IBM Corporation [14]. She is a cryptographic software developer with 15 years of experience in IBM Z Cryptography and Security. She joined IBM with a bachelor of computer science from the University of Illinois at Urbana-Champaign. After joining IBM, she continued her education with a master of information technology from Rensselaer Polytechnic Institute. Eysha is a prominent speaker for IBM Z Crypto at conferences in the USA and abroad, and has several cryptography patents [15].

Fen Zhao is an early stage investor and head of data science research at Alpha Edison [8]—a Los Angeles, CA-based venture capital firm investing in tech-driven industry transformations. Alpha Edison is focused on redesigning the machinery of cognition to reduce the noise that has overwhelmed most venture capital investment strategies and to remove cognitive biases in decision making. Prior to joining Alpha Edison, Fen Zhao developed public private partnerships at the National Science Foundation [16] in the areas of data science and cybersecurity. She created and led the Big Data Hubs and Spokes Program and was the program coordinator for the Secure and Trustworthy Cyberspace (SaTC) program. During the Obama Administration, Dr. Zhao was an AAAS Fellow at the White House Office of Science and Technology Policy (OSTP) working on national security S&T issues. Before her work in the public sector, Dr. Zhao was an associate with McKinsey and Company's Risk Management Practice, serving public sector clients with a focus on mortgage and debt markets. Fen received her PhD in Computational Astrophysics from Stanford University and her BS in Physics and Mathematics for Computer Science from MIT. Her doctoral research was conducted at the Kavli Institute for Particle Astrophysics and Cosmology at SLAC National Accelerator Labs, where she developed supercomputing astrophysical simulations of magnetic fields within the early universe. She is a native New Yorker [17].

Florence D. Hudson is founder and CEO of FDHint, LLC [18] consulting in advanced technologies and diversity & inclusion. Formerly IBM [14] vice-president and chief technology officer, Internet2 senior vice president and chief innovation officer, and an aerospace engineer at Grumman and NASA, she is special advisor for the NSF Cybersecurity Center of Excellence at Indiana University [19], and Northeast Big Data Innovation Hub at Columbia University. She serves on Boards for Princeton University, Cal Poly San Luis Obispo, Stony Brook University, and Union County College. She is cofounder of IEEE-ISTO Blockchain in Healthcare Global [20], and on the Editorial Board for Blockchain in Healthcare Today. She graduated from Princeton University with a BSE in Mechanical and Aerospace Engineering, and attended executive education at Harvard Business School and Columbia University [21].

Grace Wilson Caudill is a USAID (United States Agency for International Development) [22] scholar and an NSF EPSCoR (National Science Foundation Experimental Program to Stimulate Competitive Research Fellow). During her academic tenure she conducted research on wireless sensor motes, performing data analytics on airport runway surface conditions, and data modeling on big datasets. She attended Kentucky State University, a Historically Black College or University (HBCU) in Frankfort, Kentucky, where she earned an Associate of Science degree in Electronics Technology, two bachelor of science degrees in Computer Science and Network Engineering, and a master of science degree in Information Security and Assurance. Grace enjoys learning and engaging in bleeding-edge research, she has embarked on earning her PhD in Cybersecurity. She engages as a consultant with businesses and individuals on Cybersecurity for a Women-owned Small Business—FSS Technologies (FSST). Grace currently functions in the role of IT Auditor Principal, focusing on Cybersecurity at the University of Kentucky. She volunteers as a YMCA Certified Level II Swim Official for Kentucky and Ohio. Alongside her career and scholarly endeavors she enjoys hiking, and playing volleyball and recreational golf. Grace recently held the position of Cyberinfrastructure Engineer and Campus Champion on an NSF grant aimed at improving and supporting the cyberinfrastructure forays of researchers and students at universities. She has presented at the national level in higher education at various conferences; where she served as lead chairperson in conducting Birds-of-a-Feather (BoF) forums, leading presentations, participating in workshops and hosting scholarly discussions centered around High Performance Computing, Cyberinfrastructure, and Cybersecurity [23].

Hannah Short specializes in Trust, Identity and Security for Science. Although based at the European Organisation for Nuclear Research (CERN) [24], on the border of Switzerland and France, she spends most of her time collaborating with a network of colleagues from the Research and Education sector around the globe. After completing a master's degree in astrophysics, Hannah decided to pursue her newfound interest in programming by becoming a software developer. Since that point life has brought many and varied projects in commercial and research organizations leading to her current position at CERN. As more of our lives are spent online, security and privacy have become areas that Hannah prioritizes, for both the technical and ethical challenges. A particular topic of focus is security for distributed authentication systems, for which Hannah received a GÉANT [25] Community Award following her contribution to a trust framework for security incident response. In addition, Hannah chairs the Steering Committee for the WISE Community, Wise Information Security for collaborating E-Infrastructures [26]. WISE provides a forum for security representatives from e-Infrastructures to share best practices and, most importantly, to meet face-to-face and build trust between one another. In her home life, Hannah fills her days with mountain sports and introducing the next generation of ladies to coding—hopefully before they too reach university! Computing outreach events for female students were directly responsible for her transition into software and Hannah hopes to pay on the favor [27].

Joanna Lyn Grama, JD, CISSP, CIPT, CRISC, is a senior consultant with Vantage Technology Consulting Group [28], where she advises clients on information security policy, compliance, governance, and data privacy issues. Never content to confine her interests to a single bucket, Joanna grew interested in information security, privacy, and related legal and policy issues in higher education as a "second career." Technology development moves quickly and the law often has trouble keeping up with the pace of change. Taking advantage of job opportunities, formal and informal networks, classes, volunteer opportunities, and mentor opportunities, Joanna was thrilled to share what she learned in the textbook, LEGAL ISSUES IN INFORMATION SECURITY (Second edition, 2014). Prior to joining Vantage she was the Director of the EDUCAUSE [29] Cybersecurity and IT GRC (governance, risk, and compliance) Programs, which are designed to serve higher education IT professionals with resources, learning and professional development opportunities, and a strong peer network. She is a member of the US Department of Homeland Security's Data Privacy and Integrity Advisory Committee, appointed to committee by former Secretary Janet Napolitano, and serves as the chairperson of its technology subcommittee [30]. Joanna received her J.D. from the University of Illinois College of Law with honors and a B.A. in international relations from the University of Minnesota [31].

Karen Herrington is the director of information technology analytics and visualization at Virginia Polytechnic Institute and State University, also known as Virginia Tech [32]. An Information Technology professional with over 30 years of experience, she is a proven leader, having been at the forefront of enabling transformative technologies in both the private sector and the Higher Education arena. Karen's areas of expertise include identity management, Internet of Things, multifactor authentication, data management and analytics. She holds both a master's and a bachelor's degree in computer science from Mississippi State University [33].

Kim Milford began serving as executive director of the Research & Education Networking Information Sharing & Analysis Center (REN-ISAC) at Indiana University (IU) in 2014 [34]. She works with members, partners, sponsors, and advisory committees to direct strategic objectives in support of members, providing services and information that allow higher educational institutions to better defend local technical environments, and is responsible for overseeing administration and operations. She led an eMBA course, "Managing Information Risk and Security" for IU's Kelley School of Business. Since joining Indiana University in 2007, Ms. Milford has served in several roles leading strategic IT initiatives. As Chief Privacy Officer, she coordinated privacy-related efforts while serving on IU's Assurance Council, chairing the Committee of Data Stewards, and directing the work of the University Information Policy Office including IU's IT incident response team. From 2005 to 2007, Ms. Milford worked as Information Security Officer at the University of Rochester leading an information security program that included disaster recovery planning, identity management, incident response, and user awareness. In her position as Information Security Manager at University of Wisconsin-Madison from 1998–2005, she assisted in establishing the university's information security department and co-led in the development of an annual security conference. Ms. Milford provides cybersecurity, information policy, and privacy expertise and presentations at national and regional conferences, seminars and consortia. Ms. Milford has a B.S. in Accounting from Saint Louis University in St. Louis, Missouri and a J.D. from John Marshall Law School in Chicago, Illinois [35].

Licia Florio works for the GÉANT Association [25], as a senior trust and identity manager. Over the last 15 years, Licia has been involved in many key initiatives that make up the current European and global Authentication and Authorisation

Infrastructure for Research & Education (R&E). She supported the Task Force that produced the first eduroam (federated access to wireless networks) pilot—a service that now counts tens of thousands of hotspots in 89 countries across the world; she led the working group that created REFEDS (the Research and Education FEDerations Group) [36], the global forum that gathers R&E identity federations that now consists of 90 R&E Identity federations worldwide; she managed the European Funded project on Authentication and Authorisation for Research Collaboration (AARC) to enable federated access for large-scale research collaborations. Currently she co-leads the Trust and Identity activities in the context of the GÉANT project, the pan-European data network for the research and education community. In June 2018, Licia was awarded the prestigious Medal of Honour by the Vietsch Foundation [37] that supports research and development of advanced internet technology for scientific research and higher education [38].

Meredith M. Lee is the founding executive director of the West Big Data Innovation Hub [5], a venture launched with support from the National Science Foundation to build and strengthen partnerships across industry, academia, nonprofits, and government. Based at the University of California—Berkeley [39], Dr. Lee collaborates globally as part of a national network of Big Data Innovation Hubs to address scientific and societal challenges. Her work focuses on translational data science, including initiatives in Smart and Connected Communities, Water, Disaster Recovery, Health, and Education. She currently leads the Women in Data Science (WiDS) Datathon, a hands-on feature of the Global WiDS Conference that reached more than 100,000 participants and 150 cities in 2018 [40]. Meredith was previously an AAAS Science & Technology Policy Fellow at the Homeland Security Advanced Research Projects Agency, guiding strategic programs in graph analytics, risk assessment, machine learning, data visualization, and distributed computing. Under the Obama Administration, she led the White House Innovation for Disaster Response & Recovery Initiative and contributed to several science, technology, and open data/open government initiatives. Meredith completed her B.S., M.S., and Ph.D. in Electrical Engineering at Stanford University, and holds a US Patent from her postdoctoral research at the Canary Center for Cancer Early Detection. Her formative experiences have included working on satellite communications at MIT Lincoln Laboratory, real-time data monitoring at Agilent Laboratories, nanofabrication at IBM T.J. Watson Research Center, and microprocessors at Intel. She has taught Human-Centered Design and Innovation at Stanford as well as in government and continues to advise organizations with efforts involving technology, data science, multistakeholder collaboration, and entrepreneurship. Meredith cofounded the nonprofit NationOfMakers.org, and supports STEM efforts through serving on advisory boards including NASA DIRECT STEM, the Optical Society of America, and the National Leadership Council for the Society for Science and the Public. Her work has been featured by whitehouse.gov, ArsTechnica, The Washington Post, Forbes, WIRED, Bloomberg, and Nature [41].

Qi Pan is a digital media associate on the Future Leaders Programme at GlaxoSmithKline (GSK), a world-leading healthcare company [42]. She was the

General Data Protection Regulation (GDPR) [43] expert for Consumer Healthcare Tech, responsible for training and successfully rolling out GDPR-compliant technologies to the GSK Consumer Healthcare salesforce across EU markets in 2018. Qi organizes thought-leadership debates in GSK, where external experts are invited to bring the outside in and inspire transformative ways to benefit consumers and patients. Prior to GSK, Qi studied Molecular and Cellular Biochemistry at the University of Oxford, graduating with an MBiochem for her research into the role of epigenetics in X chromosome inactivation. Aside from work, Qi is a staunch advocator of inclusion and diversity with a focus on women in STEM [44].

Sinjoni Mukhopadhyay is a fourth-year computer science PhD candidate at the University of California, Santa Cruz [45]. Prior to beginning her PhD career, she completed her bachelor's degree in electronics and telecommunication from India, followed by her master's (Efficient Reconstruction Techniques for Disaster Recovery in Secret-Split Datastores) in computer science from the University of California, Santa Cruz. Sinjoni's areas of interest include storage security, archival storage, distributed storage, and cloud storage. She is currently working on building a self-improving synthetic workload generator using neural networks that can generate workloads which can be used to test predicted future systems. In her free time she usually reads novels, paints, or practices Indian classical dance forms like bharata natyam and odissi [46].

Soody Tronson With over 25 years of operational experience in technology, business, management, and law in start-up and fortune 100 companies, Soody's strategic insight coupled with her practical approach is her key asset, whether acting as counsel and advisor or leading her own ventures. After holding technical and management positions at Schering Plough and HP where she developed and took several products to market; and practicing law at Hewlett-Packard Inc., a successfully acquired medical device start up, and two national law firms, Heller Ehrman and Townsend and Townsend; she formed the boutique intellectual property law firm, STLGip [47], which counsels domestic and international clients in IP and technology transactions in a wide range of technologies. Soody is who you could call a Renaissance woman with a strong sense of community. She serves in board, advisory, and leadership capacities with several organizations including STEM to Market national accelerator created by the Association of Women in Science, California Lawyers Association Executive Committee of the Intellectual Property Section, Licensing Executives Society USA/Canada Women in Licensing Committee, and the Palo Alto Area Bar Association. Currently, Soody, in her role as a licensing executive, is leading efforts to provide best practices and recommendations on streamlining data sharing agreements. In this effort, she is working with the Northeast and West Big Data Innovation Hubs and the Licensing Executives Society. On the civic side, Soody is a Commissioner with the City of Menlo Park in California and an active hands-on volunteer with several civic organizations including Defy Ventures, an entrepreneurship, employment, and character development training program for currently and formerly incarcerated men, women, and youth. Soody is also founder and CEO of a consumer medical device company, Presque, developing wearable technologies to help mothers get their babies off to a healthy start. She is also the cofounder of HighNoteCoffee Co., a third-wave coffee roasting company in Silicon Valley. Soody holds a J.D., an M.S. in industrial chemistry, and a B.S. in chemistry, and is licensed to practice before the State of California and the US Patent and Trademark Office. She is a named inventor on numerous patents and patent applications covering polymer chemistry, medical devices, printing mechanisms, fluid delivery systems, sensor design, and consumer products [48].

References

1. IEEE Trust and Security Workshop for the Internet of Things, IEEE Standards Association, 4 Feb 2016. https://internetinitiative.ieee.org/images/files/events/ieee_end_to _end_trust_meeting_recap_feb17.pdf
2. IEEE Computer Society, IT Professional Magazine, Technology Solutions for the Enterprise. Enabling trust and security: TIPPSS for IoT; March/April 2018
3. IEEE Computer Society, Computer Magazine. Wearables and medical interoperability: the evolving frontier; September 2018
4. Hudson FD, Cather M (2017) TIPPSS—trust, identity, privacy, protection, safety and security for smart cities. In: Ian Abbott-Donnelly, Harold "Bud" Lawson (eds) Creating, analysing and sustaining smarter cities: a systems perspective. (ISBN-10: 1848902093, ISBN-13: 978-1848902091)
5. https://westbigdatahub.org/
6. https://www.linkedin.com/in/aliciadjohnson/
7. http://aliciadjohnson.com/
8. https://www.alphaedison.com/
9. https://www.linkedin.com/in/brittdanneman/
10. https://axoncyber.com/
11. https://www.linkedin.com/in/cynthiadiannemares/
12. https://blogs.cisco.com/author/ednaconway
13. https://www.linkedin.com/in/ednaconway/
14. https://www.ibm.com/
15. https://www.linkedin.com/in/eysha/
16. https://www.nsf.gov/
17. https://www.linkedin.com/in/fenzhao/
18. https://fdhint.com
19. https://trustedci.org/who-we-are/
20. https://www.blockchaininhealthcare.global/
21. https://www.linkedin.com/in/florencehudson/
22. https://www.usaid.gov/
23. https://www.linkedin.com/in/grace-wilson-caudill-6953873b/
24. https://home.cern/
25. https://www.geant.org/
26. https://wise-community.org/
27. https://www.linkedin.com/in/hannahshort08/
28. https://www.vantagetcg.com/
29. https://www.educause.edu/
30. https://www.dhs.gov/privacy-office-dhs-data-privacy-and-integrity-advisory-committee-membership
31. https://www.linkedin.com/in/joannagrama/
32. https://vt.edu/

33. https://www.linkedin.com/in/karen-herrington-b5aa8480/
34. https://www.ren-isac.net/
35. https://www.linkedin.com/in/kimmilford/
36. https://refeds.org/
37. http://www.vietsch-foundation.org/
38. https://www.linkedin.com/in/licia-florio-16580a2/
39. https://www.berkeley.edu/
40. https://www.widsconference.org/
41. https://www.linkedin.com/in/mmlee/
42. https://www.gsk.com/
43. https://eugdpr.org/
44. https://www.linkedin.com/in/qi-pan-19753aa0/
45. https://www.ucsc.edu/
46. https://www.linkedin.com/in/sinjoni-mukhopadhyay/
47. https://www.stlgip.com/
48. https://www.linkedin.com/in/soodytronson/

Index

A
Acceptable use policies (AUPs), 79
Adoption of IoT technology, 64
Adoption s-curve, 61
Advanced Message Queuing Protocol
    (ADMQP), 114
Alexa, 37, 40, 41, 44, 45
Algorithms, 16, 19, 28
Amazon Web Services (AWS), 162
American Standard Code for Information
    Interchange (ASCII), 146
Anti-counterfeiting chips, 10–11
Application Program Interface (API)
    framework, 94
Arizona State University (ASU), 92
Artificial intelligence (AI) techniques, 16
Asymmetric keys, 145, 146
Attribute-Based Access Control (ABAC), 136
Authentication, 87–92, 94
Authentication and Authorisation for Research
    Collaboration (AARC), 180
Authorization, 88–91, 93, 94, 135, 136
Authorization for Constrained Environments
    (ACE), 113
Automated decision making, 28, 29
Availability
    cloud computing resources, 162
    vs. cost, 171
    data, 159
    scalability, 171
    vs. vendor study, 161

B
Big Data, 131, 133, 138
Birds-of-a-Feather (BoF) forums, 178

B
Blockchain-based technology, 157
Bring Your Own Everything (BYOE)
    technology, 73, 74, 77, 80, 83, 84
Business, 133, 137, 138
Business value of trust, 62

C
California Consumer Privacy Act of 2018
    (CCPA), 122, 123
Campus Recreation Complex (CRC), 93
Cardiopulmonary resuscitation (CPR), 62
Central Intelligence Agency (CIA), 43
Central Processor Assist for Cryptographic
    Function (CPACF), 153
Centralized multi-node repair (CMR)
    model, 169
Cipher Block Chaining mode (CBC), 147, 153
Clarifying Lawful Overseas Use of Data Act
    (CLOUD), 22
Classroom policies, 80, 81
Cloud, 133, 136
The CLOUD Act, 22
Cloud based computing, 25
Compound Annual Growth Rate (CAGR), 161
Computational security, 168
Confidentiality, integrity and availability
    (CIA), 132
Connected ecosystem
    threat impacts, 5
    threats, 5
Constrained application protocol (CoAP), 114
Consumer trust in companies
    changing texture, 60–63
    customers, 57, 58
    in business, 56, 57

© Springer Nature Switzerland AG 2019
F. D. Hudson (ed.), *Women Securing the Future with TIPPSS for IoT*, Women
in Engineering and Science, https://doi.org/10.1007/978-3-030-15705-0

Consumer trust in companies (*cont.*)
    IoT, 63, 64
    market value, 58, 59
    technology, 54–56
    Uber, 53, 54
    venture capitalists, 68
Consumers, 132, 134–136, 138
Corporate board oversight, 31
Counterfeit, 5
Coupling facility (CF) encryption, 149
Credentials, 89–91
Crypto anchors, 157
Crypto Express adapter, 145, 152, 153, 156
Cryptographic keys, 144
Cryptography, 8–10, 144, 151, 153, 158
Cyber insurance, 19, 33
Cyberattacks cost, 20
Cybersecurity, 16, 21, 23, 24, 30, 32–33,
        175–177, 179
    IoT, 74–77
Cybersecurity Information Sharing Act of
        2015 (CISA), 22, 23

D
Data at-rest, 143, 150–152
Data Encryption Standard (DES), 145, 147, 151
Data governance policies, 79, 80
Data in transit, 143, 148, 149
Data privacy, 170
Data science
    and IoT, 120, 126, 127
    translational capabilities, 127
Deprovisioning, 89–91
Distributed Denial of Service (DDoS), 115
Distributed storage
    CAGR, 161
    cloud storage, 171
    consumers, 161
    existing storage technologies, 165, 166
    IoT
        devices, 164
        storage, 161–163
        security, 163, 164
    long- and short-term storage of data, 171
    Millions of Units, 161
    outline, 159, 160
    secret-sharing, 166–170

E
Electronic CodeBook mode (ECB), 147
Electronic Frontier Foundation (EFF), 123
Electronic toll collection (ETC), 64

Elliptic curve cryptography (ECC), 113, 147
Encryption, 8
    algorithm, 147
    asymmetric keys, 145, 146, 152
    in constrained environments, 9
    crypto anchors, 157
    cryptographic keys, 144
    encrypted data, 144
    hardware engines (*see* Hardware engines)
    homomorphic, 157
    in IoT environment, 8, 9
    input text, 146
    key archival, 156
    key availability, 157
    key ownership, 154
    key protection, 155
    key rotation, 156
    key storage, 154
    logical backups, 155
    management, 153, 154
    offline backups, 155
    output text, 147
    physical backups, 155
    software engines, 153
    symmetric keys, 144, 145
Engineer, 177
Equal protection clause, 28, 29
Erasure coding, 160
Ethics, 120, 124, 126
European Council for Nuclear Research
        (CERN), 97
European Economic Area (EEA), 49
European Network and Information Security
        Agency (ENISA), 24
European Organization for Nuclear Research
        (CERN), 178
Extended Binary Coded Decimal Interchange
        Code (EBCDIC), 146
eXtensible Access Control Markup Language
        (XACML), 136
Extensible Messaging and Presence Protocol
        (XMPP), 114

F
Fair Information Practice Principles
        (FIPPs), 134
Federal Trade Commission (FTC), 19, 20,
        29, 123
Federated identity and access management
        (FIAM), 108
Federated identity management (FIM), 107
    cellular phone, 111
    characteristics, 112

commercial sector, 108
data capability, 111
802.1x, 110
identity federation model, 109
institutions, 108
PKI-based access, 111
SAML, 109–111
1st Amendment, 25, 26
14th Amendment, 28, 29
4th Amendment, 27
5th Amendment, 27, 28
Format Preserving Encryption (FPE), 147

G
General Data Protection Regulation (GDPR),
        17, 18, 21, 24, 48, 49, 109, 121–123,
        127, 132, 134, 135, 142, 181
GlaxoSmithKline (GSK), 180
Global positioning system (GPS), 27
Governance, risk and compliance (GRC)
        programs, 179
Governments, 136, 137
Graphics Processing Unit (GPU), 169

H
Hardware engines
    CPU-integrated cryptographic hardware, 153
    HSMs, 152
Hardware security modules (HSMs), 152, 153
Health Information Portability and
        Accountability Act (HIPAA),
        142, 158
Health Information Technology for
        Economic and Clinical Health Act
        (HITECH), 142
Health Maintenance Organizations
        (HMOs), 142
Higher education
    adapting technology policies, 84, 85
    campus, 73, 74, 82, 83
    campus infrastructure, 77, 78
    campus policy, 78
    IoT (*see* Internet of Things (IoT))
    privacy and cybersecurity, 74–77
    resource capacity planning, 84
    risk environment, 83, 84
Historically Black College or University
        (HBCU), 177
Homomorphic encryption, 157
Human-Computer Interaction (HCI), 54
Hybrid cloud storage model, 159, 170, 171
Hypertext transfer protocol (HTTP), 165

I
IBM Crypto Express adapters, 152, 153
IBM Z systems, 154, 156
Identities
    IoT ecosystem, 89, 90
Identity, 179, 180
Identity and access management (IAM), 175
    communications, 112
    device authentication, 113, 114
    device capabilities, 112
    device connection, 114, 115
    in IoT, 111, 112
    scale, 115
    security and federated identity
        management, 112
Identity management
    authentication, 87
    demographic information, 88
    IoT, 93, 94
    university campus, 88–92
    virtual/cyber world, 87
Information, 131, 132, 134, 136–138
Information and Communications Technology
        (ICT), 5, 6, 10
Information life cycle, 142, 143
Information Sharing and Analysis Centers
        (ISAC), 23
Information Sharing and Analysis
        Organizations (ISAO), 23
Information technology (IT), 2, 5, 7, 11
Information-theoretic security, 168, 171
Infrastructure as a Service (IaaS), 163
Institute of Electrical and Electronics
        Engineers (IEEE), 1
Intellectual property (IP), 5, 7
Intellectual property rights (IPRs), 124
International Telecommunication Union
        (ITU), 112
Internet Engineering Task Force (IETF), 111
Internet of Things (IoT)
    abundance, 102–104
    in academy
        classroom policies, 80, 81
        cybersecurity hygiene, 82
        research data management policies,
            81, 82
        teaching and research, 80
    CERN, 97
    constitutional implications, 25, 26
    consumers, 132
    curiosity, 101, 102
    definition, 97
    devices, 104, 141
    digital cyberspace, 131

Internet of Things (IoT) (*cont.*)
    on earth, 97–99
    ecosystem, 1–3
    FIM, 107
    heterogeneity and resource, 134
    IAM, 116, 117
    identities, 89, 90
    incident response, 101
    medical devices, 116
    in office
        acceptable use policies, 79
        data governance policies, 79, 80
        policies, 78
    PII, 131
    research communities, 100
    resource-constrained, 135
    science, 99, 100
    secret-sharing, 170
    security, 5
    trust
        Amazon Alexa, 65, 66
        bird scooters, 65
        Tesla self-driving cars, 66–68
    users, 115, 116
Internet of Things Cybersecurity Improvement
    Act, 23
Internet protocol (IP), 7, 39, 94
IOTMDB, 165

J
Java Cryptography Extension (JCE), 153
Java Virtual Machine (JVM), 153
Judges, 175

K
Key archival, 156
Key availability, 157

L
Laboratory
    abundance of IoT, 102, 103
    "curiosity killed the cat", 101, 102
Large Hadron Collider (LHC) Design Report, 99
Laser Interferometer Gravitational Wave
    Observatory (LIGO), 100
Laws of society, 133
Lawyers, 175
Legal, 119, 124, 125, 127
Light Detection and Ranging (LiDAR)
    sensors, 26
Logical backups, 155

M
Mainframe, 146, 149, 151, 153
Manufacturers, 132
Media Access Control (MAC), 113
Medical devices, 41
Medical Research Council (MRC), 49
Medium-sized firms, 20
Message Queuing Telemetry Transport
    (MQTT), 164
MQ Telemetry Transport or Message Queuing
    Telemetry Transport (MQTT), 114
Multi-cloud strategy, 159, 160, 170, 171
Multifactor authentication, 88
Multi-secret image sharing (MSIS), 169

N
National Institute of Standards and
    Technology (NIST), 2, 8, 9, 23, 98
National Security Agency (NSA), 48
Near-field communication (NFC), 39
Network operating centers (NOCs), 77
Network security, 11
New Electronic Communications Privacy Act
    (CalECPA), 122
Nonrepudiation, 132, 133

O
OAuth2, 94
Offline backups, 155
Open ID Connect, 94
Operational security, 10
Operational Technology (OT), 2, 5, 11
Organization for the Advancement of
    Structured Information Standards
    (OASIS), 109
Owners, 135, 136

P
Panopticon, 40
Passwords, 87, 88, 91, 94
Payment Card Industry Data Security Standard
    (PCI-DSS), 142
Peripheral Component Interconnect Express
    (PCIe), 152
Personal/personally identifiable information
    (PII), 131
Pervasive encryption
    access controls, 150
    application level, 150, 151
    CF, 149
    computer memory and hardware storage, 150

computing environment, 148
database, 151
disk and tape level, 151, 152
file and data set level, 151
IoT device, 148
network, 148
Physical backups, 155
Physically Unclonable Function (PUF), 133
Platform as a Service (PaaS), 41
Policy, 175, 178, 179
Powers, E.S., 176
Privacy, 175, 178, 179
  and civil liberties guidelines, 23
  concept, 37
  consumers, 37
  and data breaches, 24
  definition, 17
  Helsinki Privacy Experiment of 2012, 49
  individual, 17
  and IoT, 74–77
    Amazon Echo, 45, 46
    cases, 40
    consumer demand, 45
    consumer industry, 41
    and data, 39, 40
    discrimination, 47
    Edge Analytics, 44
    education, 46, 47
    GDPR, 48, 49
    governmental monitoring, 48
    healthcare, 41, 42
    invasion, 38
    organizations and individuals, 48
    smart cities and crime, 43, 44
    smart homes, 42, 43
    smart offices, 43
    vast troves of data, 50
  policymakers, 38
  risks, 37
  and security, 18–20, 29, 30, 34
  surveillance capitalism, 50
Privacy management
  location, 137, 138
  privacy, 136, 137
  regulations, 133, 134
  social media and engineering, 137
Protected health information (PHI), 84, 142
Protection, 120, 122, 125, 126, 175
Provisioning, 89–91
Public Key Cryptography (PKC), 113
Public Key Infrastructure (PKI), 111
Public safety
  ethical implementation, 120
  IoT, 127

landscape, 120, 121
protections, 120, 126
and social good, 119
TIPPSS, 120

R
Radio frequency identification (RFID), 39, 43,
  48, 165
Real *vs.* Forecast Facebook Stock Price, 59
Registry, 88, 90, 93
Regulatory, 135
Reiter vs. Fairbank case, 32
Representational State Transfer (REST), 94
Research and Education FEDerations Group
  (REFEDS), 180
Research and Education Networking
  Information Sharing and Analysis
  Center (REN-ISAC), 76, 179
Research data management policies, 81, 82
Risk
  IoT systems, 134
  privacy, 131
Rivest–Shamir–Adleman (RSA), 145–147, 166
Root of trust (ROT), 133

S
Safety, 175
Sarbanes-Oxley Act of 2002, 21
Secret-sharing
  applications, 168, 169
  Internet of Things, 160, 170
  mathematics, 167, 168
  regeneration, 166
  Shamir's secret-sharing, 166
  Shamir's secret-splitting, 167
  type of, 168
Secure and Trustworthy Cyberspace (SaTC)
  program, 177
Securities and Exchange Commission (SEC),
  21, 32
Security, 175–178
  challenges, 132
  vs. compliance, 21
  Internet-enabled security cameras, 133
  privacy, 17–20, 29
  social, 132
Security architecture
  behavioral, 11
  core domains and descriptions, 7
  flexible goal-based approach, 8
  ICT third party ecosystem, 6
  IoT lifecycle, 10

Security architecture (*cont.*)
   IT network, 11
   leveraging, 6
   logical/operational, 10
   physical, 10
   prescriptive requirement, 7
Security operating centers (SOCs), 77
Sensor Web Enablement (SWE), 165
Shamir's secret-sharing scheme,
     166–169
Smart campus, 92, 93
Social engineering, 137
Social media, 137, 138
Society, 120
Software as a Service (SaaS), 41
Software cryptographic engines, 153
SURF's Cyber Threat Assessment 2017, 98
Symmetric keys, 144, 145

T
Tainted ICT, 5
Techniques, tactics and practices
    (TTPs), 74
Technology, 133, 135, 136, 138
Terabits Per Second (Tbps), 74
Third party ecosystem
   connected ecosystem, 3
     (*see* Connected ecosystem)
   coordinated deployment plan, 11
   cryptography, 8–10
   ICT, 5, 6
   IoT, 1–3, 12

   pervasive security, 6
   security architecture, 6–8, 10, 11
Third party risk, 3
Total Cost of Ownership (TCO), 161
Trust, 133, 134, 175, 176, 178, 180
Trusted Platform Module (TPM), 133
Trust, identity, privacy, protection, safety and
     security (TIPPSS), 120, 175
Trust s-curve, 61

U
Uniform Trade Secrets Act (UTSA), 126
United States Agency for International
     Development (USAID), 177
United States vs. Jones case, 27
University campus, 88–94
US Constitution, 15, 25, 27, 28
User-Managed Access (UMA), 94

V
Value chain, 7
Value sensitive design (VSD), 54, 55
Vehicle-to-vehicle (V2V) communication, 163

W
White House Office of Science and
     Technology Policy (OSTP), 177
Wireless Sensor Network (WSN), 120, 133
Women in Data Science (WiDS), 180
Women in STEM, 181

Printed in the United States
By Bookmasters